NUREG-1814, Rev. 1

# Status of the Decommissioning Program

# 2006 Annual Report

# Final Report

Manuscript Completed: December 2006
Date Published: February 2007

Prepared by
J. Buckley

Division of Waste Management and Environmental Protection
Office of Federal and State Materials
and Environmental Management Programs
U.S. Nuclear Regulatory Commission
Washington, DC 20555-0001

## ABSTRACT

This report provides a comprehensive overview of the U.S. Nuclear Regulatory Commission's (NRC's) decommissioning program. Its purpose is to provide a stand-alone reference document that describes the decommissioning process and summarizes the status of decommissioning activities, under NRC jurisdiction, through September 30, 2006. This includes the decommissioning of complex decommissioning sites, commercial reactors, research and test reactors, uranium recovery facilities, and fuel cycle facilities. In addition, this report discusses accomplishments of the decommissioning program in fiscal year (FY) 2006; identifies the key decommissioning program issues that the staff will address in FY 2007; and provides information Agreement States have supplied on decommissioning in their States.[1]

## PAPERWORK REDUCTION ACT STATEMENT

The information collections contained in this NUREG are covered by the requirements of Title 10 of the Code of Federal Regulations, Parts 19, 20, 30, 33, 34, 35, 36, 39, 40, 51, 70, 72, and 150, which the Office of Management and Budget (OMB) has approved, approval numbers 3150-0044, 0014, 0017, 0015, 0007, 0010, 0158, 0130, 0020, 0021, 0009, 0132, and 0032.

## PUBLIC PROTECTION NOTIFICATION

If a means used to impose an information collection does not display a currently valid OMB control number, NRC may not conduct or sponsor, and a person is not required to respond to, the information collection.

---

[1]As of September 2006, NRC is planning to reorganize the Office of Nuclear Material Safety and Safeguards (NMSS) and the Office of State and Tribal Programs (STP), to create two new offices: the Office of Federal and State Materials and Environmental Management Programs (FSME), which will focus on materials programs; and the new NMSS, which will focus on fuel cycle programs. This reorganization is scheduled to take effect on October 1, 2006. This document contains references to NMSS and STP. These references will be updated in future revisions of this document.

# TABLE OF CONTENTS

**APPENDICES**

**TABLES**

## ACKNOWLEDGMENTS

This report is a compilation of information from several U.S. Nuclear Regulatory Commission Offices. As such, many individuals provided assistance in the development and review of this NUREG report.

Participants:

Larry Camper
Claudia Craig
Patrick Isaac
Robert Johnson
Keith McConnell
Dominick Orlando
William Ott
Stewart Treby
Stephen J. Cohen
Erikka Legrand
William Rautzen

## ABBREVIATIONS

| | |
|---|---|
| ACL | alternate concentration limit |
| AEA | Atomic Energy Act of 1954 |
| AEC | U.S. Atomic Energy Commission |
| ASLBP | Atomic Safety and Licensing Board Panel |
| ATK | Alliant Ordinance and Ground Systems, LLC |
| BTP | branch technical position |
| CERCLA | Comprehensive Environmental Response, Compensation, and Liability Act of 1980 |
| CFR | *Code of Federal Regulations* |
| Ci | curie |
| Co | cobalt |
| CRADAL | Computerized Risk Assessment and Data Analysis Lab |
| CRCPD | Conference of Radiation Control Program Directors |
| Cs | cesium |
| CSM | Consultant Services Meeting |
| CY | calendar year |
| DCGL | Derived Concentration Guideline Level |
| DGSNR | Directorate General for Nuclear Safety Agency |
| DOE | U.S. Department of Energy |
| DP | Decommissioning Plan |
| DU | depleted uranium |
| DWMEP | Division of Waste Management and Environmental Protection |
| EA | environmental assessment |
| EIS | environmental impact statement |
| EPA | U.S. Environmental Protection Agency |
| FA | financial assurance |
| FCSS | Division of Fuel Cycle Safety and Safeguards |
| FONSI | finding of no significant impact |

| | |
|---|---|
| FSME | Office of Federal and State Materials and Environmental Management Programs |
| FSS | final status survey |
| FSSP | Final Status Survey Plan |
| FSSR | Final Status Survey Report |
| FTE | full-time equivalents |
| FUSRAP | Formerly Utilized Sites Remedial Action Program |
| FY | fiscal year |
| g | gram |
| GPS | groundwater protection standard |
| GSA | U.S. General Services Administration |
| IC | institutional control |
| IAEA | International Atomic Energy Agency |
| IDIP | Integrated Decommissioning Improvement Plan |
| ISCMEM | Interagency Steering Committee on Multimedia Environmental Models |
| ISCORS | Interagency Steering Committee on Radiation Standards |
| ISFSI | independent spent fuel storage installation |
| ISL | in situ leach |
| km | kilometers |
| kW | kilowatt |
| l | liter |
| LLW | low-level waste |
| LTP | License Termination Plan |
| LTR | License Termination Rule |
| LTSP | Long-Term Surveillance Plan |
| MARLAP | Multi-Agency Radiological Laboratory Analytical Protocols |
| MARSSIM | Multi-Agency Radiation Survey and Site Investigation Manual |
| MDA | minimum detectable activity |
| MDEQ | Michigan Department of Environmental Quality |

| | |
|---|---|
| MED-AEC | Manhattan Engineering District and the Atomic Energy Commission |
| MOU | memorandum of understanding |
| mrem | millirem |
| NEPA | National Environmental Policy Act |
| NMSS | Office of Nuclear Material Safety and Safeguards |
| NPL | National Priorities List |
| NRC | U.S. Nuclear Regulatory Commission |
| NRR | Office of Nuclear Reactor Regulation |
| OAS | Organization of Agreement States |
| ODEQ | Oklahoma Department of Environmental Quality |
| OECD | Organization for Economic Cooperation and Development |
| OGC | Office of the General Counsel |
| OIP | Office of International Programs |
| OMB | Office of Management and Budget |
| ORISE | Oak Ridge Institute for Science and Education |
| ORNL | Oak Ridge National Laboratory |
| PA | preliminary assessment |
| PADEP | Pennsylvania Department of Environmental Protection |
| PART | Program Assessment Rating Tool |
| POTW | Publicly Owned Treatment Works |
| PSDAR | Post-Shutdown Decommissioning Activities Report |
| Pu | plutonium |
| R&D | research and development |
| RAI | request for additional information |
| RES | Office of Nuclear Regulatory Research |
| RIS | Regulatory Issue Summary |
| ROD | Record of Decision |
| RP | Remediation Plan |

| | |
|---|---|
| SDMP | Site Decommissioning Management Plan |
| SER | Safety Evaluation Report |
| SNM | special nuclear material |
| Sr | strontium |
| SRM | staff requirements memorandum |
| SRP | Standard Review Plan |
| STP | Office of State and Tribal Programs |
| TAG | Technical Advisory Group |
| TBD | to be determined |
| Tc | technetium |
| Th | thorium |
| TS | technical specification |
| U | uranium |
| UMTRCA | Uranium Mill Tailings Radiation Control Act |
| USACE | U.S. Army Corps of Engineers |
| WCS | Waste Control Specialists |
| WDEQ | Wyoming Department of Environmental Quality |
| WPDD | Working Party on Decommissioning and Dismantlement |
| WASSC | Waste Safety Standards Committee |
| WVDP | West Valley Demonstration Project |
| yr | year |

# ALPHABETICAL LISTING OF SITE SUMMARIES BY SITE CATEGORY

## SITE SUMMARIES FOR DECOMMISSIONING POWER REACTORS

## SITE SUMMARIES FOR RESEARCH AND TEST REACTORS

## SITE SUMMARIES FOR COMPLEX DECOMMISSIONING SITES

## SITE SUMMARIES FOR DECOMMISSIONING TITLE II SITES

## SITE SUMMARIES FOR DECOMMISSIONING FUEL CYCLE FACILITIES

# 1.    Introduction

This report provides a comprehensive summary of the U.S. Nuclear Regulatory Commission's (NRC's) decommissioning program. Its purpose is to provide a reference document that summarizes the decommissioning activities in fiscal year (FY) 2006, including the decommissioning of complex material sites, commercial reactors, research and test reactors, uranium mill tailings facilities, and fuel cycle facilities. In addition, this report discusses accomplishments of the decommissioning program since last year's report, provides information supplied by Agreement States on decommissioning in their States, and identifies key decommissioning program issues that the staff will address in the coming year.

# 2.    Decommissioning Sites

NRC regulates the decontamination and decommissioning of materials and fuel cycle facilities, power reactors, research and test reactors, and uranium recovery facilities. The purpose of the decommissioning program is to ensure that NRC-licensed sites, and sites that were or could be licensed by NRC, are decommissioned in a safe, timely, and effective manner so that they can be returned to beneficial use and that stakeholders are informed and involved in the process, as appropriate. A broad spectrum of activities associated with these program functions is summarized in this report.

On June 17, 2004, the elimination of the Site Decommissioning Management Plan (SDMP) designation was announced in the *Federal Register* (69 *Federal Register* 33946). NRC now manages materials decommissioning sites as "complex sites," under a comprehensive decommissioning program. The SDMP designation will be used in this report only to describe the cleanup criteria before the License Termination Rule (LTR).

Approximately 200 materials licenses are terminated each year. Most of these license terminations are routine, and the sites require little, if any, remediation to meet NRC's unrestricted release criteria. The decommissioning program discussed in this report focuses on the termination of licenses that are not routine, because the sites involve more complex decommissioning activities.

There are 16 nuclear power reactors, 14 research and test reactors, 32 complex decommissioning materials facilities, three fuel cycle facilities (partial decommissioning), and 12 uranium recovery facilities that are undergoing non-routine decommissioning or are in long-term safe storage, under NRC jurisdiction. Appendices A - E of this report contain site status summaries for the facilities managed under the decommissioning program. These summaries describe the status of each site and identify the current technical and regulatory issues impacting completion of decommissioning. For those licensees that have submitted a decommissioning plan (DP) or license termination plan (LTP), the schedules are based on an assessment of the complexity of the DP or LTP review. For those licensees that have not submitted a DP or LTP, the schedules are based on other licensee information available, and the anticipated decommissioning approach.

Through the Agreement State Program, 34 States have signed formal agreements with NRC, by which those States have assumed regulatory responsibility over certain byproduct, source, and small quantities of special nuclear material, including the decommissioning of some complex materials sites. Agreement States do not have regulatory authority over operating or decommissioning nuclear power plants. NRC's coordination with the Agreement States decommissioning programs is discussed in more detail in Section 7 of this report.

## 2.1 Nuclear Power Reactor Decommissioning

In FY 2006, NMSS had regulatory project management responsibility for 12 decommissioning power reactors. The Office of Nuclear Reactor Regulation (NRR) had project management responsibility for two decommissioning reactors (Indian Point – Unit 1, Millstone – Unit 1). In addition, NRR had project management for two decommissioning early demonstration reactors—Vallecitos, and the Nuclear Ship Savannah. Table 2–1 identifies the power reactors undergoing decommissioning.

On October 1, 2006, project management and oversight responsibility for Indian Point – Unit 1, Millstone – Unit 1, Vallecitos, and the Nuclear Ship Savannah will transfer from NRR to FSME. Plant status summaries for all decommissioning reactors are provided in Appendix A.

### 2.1.1 Decommissioning Process

The decommissioning process begins when a licensee decides to permanently cease operations. Several major steps make up the decommissioning process: notification; submittal and review of the Post-Shutdown Decommissioning Activities Report (PSDAR); submittal and review of the LTP; implementation of the LTP; and completion of decommissioning.

### Notification

When the licensee has decided to permanently cease operations, it is required to submit a written notification to NRC. In addition, the licensee is required to notify NRC in writing once fuel has been permanently removed from the reactor vessel.

### Post-Shutdown Decommissioning Activities Report

Before, or within 2 years after cessation of operations, the licensee must submit a PSDAR. The PSDAR must include:

- A description and schedule for the planned decommissioning activities;
- An estimate of the expected costs; and
- A discussion that provides the means for concluding that the environmental impacts associated with decommissioning activities will be bounded by appropriately issued environmental impact statements (EIS').

NRC will notice receipt of the PSDAR in the *Federal Register* and make the PSDAR available for public comment. In addition, NRC will hold a public meeting in the vicinity of the licensee's facility, to discuss the PSDAR. NRC does not approve the PSDAR.

The licensee cannot perform any major decommissioning activities until 90 days after NRC has received the PSDAR. After this period, the licensee can perform decommissioning activities as long as the activities do not:

- Foreclose release of the site for unrestricted use;

- Result in significant environmental impacts not previously reviewed; or

- Result in there no longer being reasonable assurance that adequate funds will be available for decommissioning.

In taking actions permitted under Title 10 of the Code of Federal Regulations (CFR) 50.59, after submittal of the PSDAR, the licensee must notify NRC, in writing, before performing any decommissioning activity inconsistent with, or making any significant schedule change from, those actions and schedules in the PSDAR.

## LTP

Each power reactor must submit an application for termination of its license. The application must be accompanied or preceded by an LTP submitted for NRC approval. The NRC and licensee generally hold presubmittal meetings to agree on the format and content of the LTP. These meetings are intended to improve the efficiency of the LTP development and review process. The LTP must include:

- A site characterization;

- Identification of remaining dismantlement activities;

- Plans for site remediation;

- Detailed plans for the final radiation survey;

- A description of the end use of the site, if restricted;

- An updated site-specific estimate of remaining decommissioning costs; and

- A supplement to the environmental report describing any new information or significant environmental change associated with the licensee's proposed termination activities.

In addition, the licensee must demonstrate that the applicable requirements of the LTR will be met.

NRC will notice receipt of the LTP and make the LTP available for public comment. In addition, NRC will hold a public meeting in the vicinity of the licensee's facility to discuss the LTP and the LTP review process. The technical review is guided by NUREG-1700, "Standard Review Plan for Evaluating Nuclear Power Reactor License Termination Plans." The LTP is approved by license amendment.

## Implementation of the LTP

NRC staff will inspect the licensee during decommissioning operations to ensure compliance with the LTP. These inspections will normally include in-process and confirmatory radiological surveys.

Decommissioning must be completed within 60 years of permanent cessation of operations, unless otherwise approved by the Commission.

Completion of Decommissioning

At the conclusion of decommissioning activities, the licensee will submit a final status survey report (FSSR) which identifies the final radiological conditions of the site and requests that NRC either: (1) terminate the 10 CFR Part 50 license; or (2) reduce the Part 50 license boundary to the footprint of the ISFSI. For decommissioning reactors with no ISFSI or an ISFSI licensed under 10 CFR Part 72, completion of decommissioning will result in the termination of the Part 50 license. For reactors with an ISFSI licensed under the provisions of Part 50, completion of decommissioning will result in reducing the Part 50 license boundary to the footprint of the independent spent fuel storage installation (ISFSI). NRC will approve the FSSR and licensee's request if it determines that:

- The remaining dismantlement has been performed in accordance with the approved LTP; and

- The final radiation survey and associated documentation demonstrate that the facility and site are suitable for release in accordance with the LTR.

2.1.2   Summary of FY 2006 Activities

NRC power reactor decommissioning activities include: (a) project management for decommissioning power reactors and technical review responsibility for licensee submittals in support of decommissioning; (b) core inspections; (c) supporting development of rulemaking and guidance; (d) conducting public outreach, including the development of communication plans; and, (e) participating in industry conferences and workshops.

- During the past year, NRR terminated Saxton's Part 50 license. NRC also approved the release of non-impacted areas from the Yankee Rowe Part 50 License. Table 2–1 provides a schedule for reactor decommissioning activities.

- One of the goals identified in NRC's Strategic Plan is to ensure openness in its regulatory process. The Strategic Plan identifies the development of communication plans for specific activities associated with the regulation of radiological decommissioning, as a means to support the openness strategy. The staff continues to implement communication plans for all decommissioning reactors. Site-specific communication plans are useful tools to help ensure that the appropriate stakeholders are identified and contacted and focuses the staff on messages NRC wants to convey.

- The staff also participated in a number of industry conferences and workshops. Examples of conferences and workshops attended by the staff during the past year include: (1) Waste Management '06; (2) American Nuclear Society conferences; (3) Fuel Cycle Facility Forum materials site meeting; (4) International Conference on Environmental Remediation and Radioactive Waste Management 05; and (5) an Electric Power Research Institute (EPRI) workshop.

## 2.2    Research and Test Reactor Decommissioning

NRR provides project management and inspection oversight for 14 decommissioning research and test reactors. As of September 30, 2006, 10 research and test reactors have decommissioning orders or amendments. Additionally, three research and test reactors are in "possession-only" status, either waiting for shutdown of another research or test reactor at the site, or for removal of the fuel from the site by the U.S. Department of Energy (DOE), and one decommissioning amendment request is under review.

On October 1, 2006, project management and oversight responsibility for the decommissioning research and test reactors will transfer from NRR to FSME. Plant status summaries for the decommissioning research and test reactors are provided in Appendix B.

### 2.2.1   Decommissioning Process

In general, the decommissioning process for research and test reactors and power reactors is the same (see Section 2.1.1).

### 2.2.2   Summary of FY 2006 Activities

In FY 2006, NRR terminated the licenses of three research and test reactors: Manhattan College, University of Virginia, and University of Virginia - Cavalier. In addition, NRC participated in the September 2006, Test Research and Training Reactors conference held in Austin, Texas. Table 2–2 identifies the decommissioning research and test reactors and provides the current status.

## Table 2-1
### Power Reactors Undergoing Decommissioning

| | Reactor | Location | PSDAR* Submitted | LTP Submitted | LTP Approved | Completion of Decomm.** | Site Summ. Pg. No. |
|---|---|---|---|---|---|---|---|
| 1 | Big Rock Point | Charlevoix, MI | 9/97 | 4/03 | 3/05 | 2007 | Page A-1 |
| 2 | Dresden – Unit 1 | Dresden, IL | 6/98 | TBD | TBD | TBD | Page A-3 |
| 3 | Fermi – Unit 1 | Newport, MI | 4/98 | 2007*** | TBD | 2008 | Page A-5 |
| 4 | Haddam Neck – Connecticut Yankee | Meriden, CT | 8/97 | 7/00 | 11/02 | 2007 | Page A-6 |
| 5 | Humboldt Bay | Eureka, CA | 2/98 | 2009*** | TBD | 2011 | Page A-7 |
| 6 | Indian Point – Unit 1 | Buchanan, NY | 1/96 | TBD | TBD | TBD | Page A-9 |
| 7 | La Crosse | La Crosse, WI | 5/91 | TBD | TBD | TBD | Page A-10 |
| 8 | Millstone – Unit 1 | Waterford, CT | 6/99 | TBD | TBD | TBD | Page A-11 |
| 9 | Nuclear Ship Savannah | Newport News, VA | TBD | TBD | TBD | TBD | Page A-12 |
| 10 | Peach Bottom – Unit 1 | Delta, PA | 6/98 | 2012*** | TBD | 2014 | Page A-13 |
| 11 | Rancho Seco | Sacramento, CA | 12/94 | 4/06 | 2007*** | 2008 | Page A-14 |
| 12 | San Onofre – Unit 1 | San Clemente, CA | 12/98 | TBD | TBD | 2045 | Page A-15 |
| 13 | Three Mile Island – Unit 2 | Harrisburg, PA | 2/79 | TBD | TBD | 2014 | Page A-17 |
| 14 | Vallecitos | Pleasanton, CA | 7/66 | TBD | TBD | TBD | Page A-18 |
| 15 | Yankee Rowe | Greenfield, MA | 11/94 | 11/03 | 4/05 | 2007 | Page A-19 |

**Table 2–1**
**Power Reactors Undergoing Decommissioning**

| | Reactor | Location | PSDAR* Submitted | LTP Submitted | LTP Approved | Completion of Decomm.** | Site Summ. Pg. No. |
|---|---|---|---|---|---|---|---|
| 16 | Zion – Units 1 & 2 | Waukegan, IL | 2/00 | TBD | TBD | 2026 | Page A-20 |

\* Post-shutdown Decommissioning Activities Report (PSDAR) or decommissioning plan (DP) equivalent.

\*\* For decommissioning reactors with no ISFSI or an ISFSI licensed under Part 72, completion of decommissioning will result in the termination of the Part 50 license. For reactors with an ISFSI licensed under the provisions of Part 50, completion of decommissioning will result in reducing the Part 50 license boundary to the footprint of the ISFSI.

\*\*\* Estimated date.

NOTE: Licensees submitted DPs (or equivalent) before 1996 and PSDARs after 1996.

**Table 2-2**

**Research and Test Reactors Undergoing Decommissioning**

| | Reactor | Location | Status | Completion of Decomm. | Site Summ. Pg. No. |
|---|---|---|---|---|---|
| 1 | Cornell University – TRIGA | Ithica, NY | DECON Approved | 2007 | Page B-1 |
| 2 | Cornell University – ZPR | Ithica, NY | DECON Approved | 2007 | Page B-2 |
| 3 | Ford Nuclear Reactor | Ann Arbor, MI | DECON Approved | TBD | Page B-3 |
| 4 | General Atomics – TRIGA Mark F | San Diego, CA | DECON Approved | TBD | Page B-4 |
| 5 | General Atomics – TRIGA Mark I | San Diego, CA | DECON Approved | TBD | Page B-5 |
| 6 | General Electric Co. – GETR | Sunol, CA | Possession-Only | TBD | Page B-6 |
| 7 | General Electric Co. – VESR | Sunol, CA | Possession-Only | TBD | Page B-7 |
| 8 | NASA - Mockup | Sandusky, OH | DECON Approved | 2010 | Page B-8 |
| 9 | NASA - Plum Brook | Sandusky, OH | DECON Approved | 2010 | Page B-9 |
| 10 | University of Buffalo | Buffalo, NY | Possession-Only | TBD | Page B-10 |
| 11 | University of Illinois | Urbana, IL | DECON Approved | TBD | Page B-11 |
| 12 | University of Washington | Seattle, WA | DECON Approved | 2007 | Page B-12 |
| 13 | Veterans Administration | Omaha, NE | DECON Amendment | TBD | Page B-13 |
| 14 | Westinghouse | New Stanton, PA | DECON Approved | TBD | Page B-14 |

Note: DECON - decontamination; GETR - General Electric Test Reactor; NASA - National Aeronautics and Space Administration; TBD - to be determined; TRIGA - Training, Research, Isotopes General Atomics; VESR - Vallecitos Experimental Superheat Reactor; ZPR - Zero Power Reactor.

## 2.3    Complex Material Facility Decommissioning

There are 32 complex materials sites undergoing decommissioning (see Table 2-3).  Table 2-3 identifies the clean-up criteria for each complex site as either LTR or SDMP Action Plan criteria. The LTR (10 CFR Part 20, Subpart E) authorized two different sets of cleanup criteria—the concentration-based SDMP Action Plan criteria and the dose-based LTR criteria.  Under the provisions of 10 CFR 20.1401(b), any licensee that submitted its DP before August 20, 1998, and received NRC approval of that DP before August 20, 1999, could use the SDMP Action Plan criteria for site remediation.  In the SRM on SECY-99-195, the Commission granted an extension of the DP approval deadline, for 12 sites, to August 20, 2000.  In September 2000, the staff notified the Commission that all 12 DPs were approved by the deadline.  All other sites must use the dose-based criteria of the LTR.

NRC has eliminated the SDMP designation for certain decommissioning facilities.  Instead, NRC manages all materials decommissioning sites as "complex sites," under a comprehensive decommissioning program.  The SDMP designation will be used in this paper only to describe decommissioning activities that have taken place before June 17, 2004.

Status summaries for the Complex Materials Sites undergoing decommissioning are provided in Appendix C.

### 2.3.1    Decommissioning Process

The decommissioning process is initiated by any one of the following conditions:

- The license expires;
- The licensee has decided to permanently cease principal activities at the entire site or in any separate building or outdoor area;
- No principal activities have been conducted for a period of 24 months; or
- No principal activities have been conducted for a period of 24 months in any separate building or outdoor area.

Several major steps make up the decommissioning process:  notification; submittal and review of the DP; implementation of the DP; and completion of decommissioning.

Notification

Within 60 days of the occurrence of any of the triggering conditions, the licensee is required to notify NRC of such occurrence and either begin decommissioning or, if required, submit a DP within 12 months of notification and begin decommissioning after approval of the plan. Alternative schedules are authorized under the regulations, with NRC approval.

## DP

A DP must be submitted if required by license condition or if the procedures and activities necessary to decommission have not been previously approved by NRC and the procedures could increase potential health and safety impacts on workers or the public, such as in any of the following cases:

• Procedures would involve techniques not applied routinely during clean-up or maintenance operations;

• Workers would be entering areas not normally occupied where surface contamination and radiation levels are significantly higher than routinely encountered during operation;

• Procedures could result in significantly greater airborne concentrations than are present during operations; or

• Procedures could result in significantly greater releases of radioactive material to the environment than those associated with operations.

Before submitting a DP, it is generally useful for the licensee to meet with NRC, to agree on the form and content of the DP. This pre-submittal meeting is intended to make the DP review process more efficient by reducing the need for requests for additional information (RAIs).

The DP review process begins with an acceptance review. Although primarily an administrative review, the acceptance review includes, but is not be limited to: (a) completeness of the application; (b) legibility of drawings; (c) general adequacy of information; (d) justification for proprietary information; and (e) obvious technical inadequacies. The objective of the acceptance review is to verify that the application contains sufficient information before the staff begins an in-depth technical review. In addition, a limited technical review will be conducted. The purpose of the limited technical review is to identify significant technical deficiencies at an early stage, thereby precluding a detailed technical review of a technically incomplete submittal. At the conclusion of the acceptance review, the DP will either be accepted for detailed technical review or rejected and returned to the licensee with the deficiencies identified. For DPs proposing unrestricted release, a full technical review will be initiated after the successful conclusion of the acceptance review. The staff's review is guided by NUREG-1757, "Consolidated NMSS Decommissioning Guidance," and its supporting references. The results of the staff's review will be documented in an Environmental Assessment (EA) and a Safety Evaluation Report (SER). The EA will be shared with the appropriate State, and State comments will be considered in finalizing the EA. The final EA must be summarized in the *Federal Register* in the form of a Finding of No Significant Impact (FONSI) unless there was a significant environmental impact in which case an EIS would be required.

For reviews of DPs proposing restricted release, the review will be conducted in two phases. The first phase of the review will focus on the financial assurance (FA) and institutional control (IC) provisions of the DP. The review of the remainder of the DP will be initiated only after the staff is satisfied that the licensee's proposed IC & FA provisions will comply with the requirements of the LTR (Part 20, Subpart E). The applicable portions of NUREG-1757, "Consolidated NMSS Decommissioning Guidance," will be used to guide this phase of the review. Phase II of the review will address all other sections of the technical review as guided by NUREG-1757 and will include the development of an EIS. Therefore, one of the first steps in Phase II is the publication of a Notice of Intent to develop an EIS. The basic EIS development steps are:

- Notice of Intent;

- Public scoping meeting;

- Preparation and publication of the scoping report;

- Preparation and publication of the draft EIS;

- Public comment period on the draft EIS, including a public meeting;

- Preparation and publication of the final EIS; and

- Preparation and publication of the Record of Decision (ROD).

In parallel with the development of the EIS, the staff will develop a draft and final SER. The development of the draft SER will be coordinated with the development of the draft EIS so that any RAIs can be consolidated.

Regardless of whether an EA or EIS is developed, the staff structures its reviews so that the number of RAIs is minimized, without diminishing the technical quality or completeness of the licensee's ultimate submittal. For example, the staff will first develop a set of additional information needs and clarifications, including the bases for the additional information/ clarifications, and then meet with the licensee or responsible party to discuss the issues. This meeting will be noticed and conducted in accordance with NRC requirements for meetings open to the public. The results of the meeting will be documented in a meeting report. Any issues that cannot be resolved during the meeting will be included in the formal RAI. In developing the final RAI, staff will document the insufficient or inadequate information submitted by the licensee and communicate what additional information is needed to address the identified deficiencies.

After publication of the FONSI (for a DP involving an EA) or the ROD (for a DP involving an EIS), a license amendment will be issued, approving the DP, along with any additional license conditions found to be necessary as a result in the EA/EIS and/or the SER.

Implementation of the DP

After approval of the DP, the licensee must complete decommissioning in accordance with the approved DP within 24 months or apply for an alternate schedule. NRC staff will inspect the licensee during decommissioning operations to ensure compliance with the DP. These inspections will normally include in-process and confirmatory radiological surveys.

Completion of Decommissioning

As the final step in decommissioning, the licensee is required to:

- Certify the disposition of all licensed material, including accumulated wastes, by submitting a completed NRC Form 314 or equivalent information; and

11

- Conduct a radiation survey of the premises where licensed activities were carried out (in accordance with the procedures in the approved DP, if a DP is required) and submit a report of the results of the survey, unless the licensee demonstrates in some other manner that the premises are suitable for release in accordance with the LTR.

Licenses are terminated by written notice to the licensee when NRC determines that:

- Licensed material has been properly disposed of;

- Reasonable effort has been made to eliminate residual radioactive contamination, if present;

- Site meets the approved DP; and

- Radiation survey has been performed or other information submitted by the licensee that demonstrates that the premises are suitable for release in accordance with the LTR.

## 2.3.2   Summary of FY 2006 Activities

Material facilities decommissioning activities include:  (a) maintaining regulatory oversight of complex decommissioning sites; (b) undertaking financial assurance reviews; (c) examining issues and funding options to facilitate remediation of sites in non-Agreement States; (d) interacting with the U.S. Environmental Protection Agency (EPA) and Interagency Steering Committee on Radiation Standards (ISCORS); (e) inspecting complex decommissioning sites; (f) conducting public outreach; (g) participating in international decommissioning activities; (h) conducting a program evaluation; and (i) participating in industry conferences and workshops.

- Since last year's status report, seven sites were removed from the complex site list through license termination or completion of decommissioning:  (1) Department of the Army - Ft. Belvoir; (2) Dow Chemical; (3) Kerr McGee Cushing; (4) Kirtland Air Force Base; (5) Heritage Minerals; (6) Union Carbide Corporation; and (7) Westinghouse Electric - Blairsville.

- Activities associated with complex site decommissioning program include:  (a) review and approval of DPs; (b) conduct of pre-DP development meetings with licensees; (c) review of licensee FSSRs and conduct of confirmatory surveys; (d) conduct of in-process inspections; and (e) preparation of EAs and SERs.  In FY 2006, the staff approved DPs for Dow Chemical Co., and S.C. Holdings, Inc.  The staff is currently reviewing DPs for the following sites: (a) AAR; (b) Cabot Corporation; (c) Curtis-Wright Cheswick; (d) Mallinckrodt Chemical, Inc.; (e) Quehanna; (f) Westinghouse Electric Company (Hematite Facility); (g) Shieldalloy Metallurgical Corp.; and (h) UNC Naval Products.

- Staff routinely reviews financial assurance submittals for materials and fuel cycle facilities, and maintains a financial instrument security program.  Approximately 25 financial assurance submittals were reviewed in FY 2006, including two complex reviews for fuel enrichment license applications.

- One of the goals identified in NRC's Strategic Plan is to ensure openness in its regulatory process.  The Strategic Plan identifies the development of communication plans for specific activities associated with the regulation of radiological decommissioning, as a means to support the openness strategy. The staff continues to implement communication plans for all complex sites. Site-specific communication plans are useful tools to help ensure that the appropriate stakeholders are identified and contacted and focuses the staff on messages NRC wants to convey. One of the activities identified in the communication plans for each site is participation in public meetings to inform the public about major licensing actions. During the past year, the staff participated in public meetings for the: (a) West Valley Demonstration Project (WVDP); (b) Mallinckrodt Chemical, Inc.; (c) Michigan Department of Natural Resources; (d) S.C. Holdings Inc.; (e) Quehanna; (f) Heritage Minerals Inc.; and (g) Pathfinder.

• The staff's participation in International activities is discussed in Section 5.

• The staff also participated in a number of industry conferences and workshops. Examples of conferences and workshops attended by the staff during the past year include: (1) Waste Management '06; (2) American Nuclear Society conferences; (3) Fuel Cycle Facility Forum meetings; (4) ICEM 05; and (5) an EPRI workshop.

## 2.4    Uranium Recovery Facility Decommissioning

Currently, 12 uranium recovery facilities are in decommissioning. Uranium recovery decommissioning activities in the Division of Fuel Cycle Safety and Safeguards (FCSS) include: (a) regulatory oversight of decommissioning uranium recovery (milling) sites; (b) review of site characterization plans and data; (c) review and approval of reclamation plans (RPs); (d) preparation of EAs and EIS'; (e) inspection of decommissioning activities, including confirmatory surveys; (f) decommissioning cost estimate reviews (including annual surety updates); and (g) oversight of license termination. Regulations governing uranium recovery facility decommissioning are found in 10 CFR Part 40 and Part 40, Appendix A. These licensees include conventional uranium mills and in-situ leach (ISL) facilities. Table 2–4 identifies the Title II decommissioning sites. Site status summaries for each of the facilities are provided in Appendix D.

On October 1, 2006, responsibility for uranium recovery decommissioning activities will be transferred from the FCSS to the Division of Waste Management and Environmental Protection (DWMEP), as part of the consolidation of NRC's decommissioning program. Details of the consolidation are provided in SECY-06-0106.

**Table 2-3**
**Complex Decommissioning Sites**

| | Name | Location | Date DP Submitted | Date DP Approved | Cleanup Criteria | Projected Removal | Site Summ. Pg. No. |
|---|---|---|---|---|---|---|---|
| 1 | AAR Manufacturing, Inc. | Livonia, MI | 10/97 revised 9/06 | 5/98 TBD | LTR-RES | 9/08 | Page C-1 |
| 2 | ABB Prospects, Inc. | Windsor, CT | 4/03 | 6/04 | LTR-UNRES | 12/07 | Page C-2 |
| 3 | Babcock & Wilcox (Shallow Land Disposal Area) | Vandergrift, PA | 6/01 revised NA | TBD | LTR-UNRES | 10/09 | Page C-4 |
| 4 | Battelle Columbus Laboratories | Columbus, OH | 8/00 | 2001 | LTR-UNRES | 11/06 | Page C-6 |
| 5 | Cabot Performance Materials, Inc. | Reading, PA | 6/05 revised 8/06 | 2/07* | LTR-UNRES | 10/07 | Page C-8 |
| 6 | Curtis-Wright Cheswick | Cheswick, PA | 3/06 | 6/07 | LTR-UNRES | 12/08 | Page C-10 |
| 7 | Department of the Army (Ft. McClellan) | Fort McClellan, AL | 3/99 | 3/01 | LTR-UNRES | 12/06 | Page C-11 |
| 8 | Eglin Air Force Base | Walton County, FL | 8/03 | 9/05 | LTR-UNRES | 12/06 | Page C-12 |
| 9 | Engelhard Minerals | Great Lakes, IL | NA | NA | LTR-UNRES | TBD | Page C-13 |
| 10 | FMRI (Fansteel) Inc. | Muskogee, OK | 8/99 Revised 5/03 | 12/03 | LTR-UNRES | 2023** | Page C-15 |

14

**Table 2-3**
**Complex Decommissioning Sites**

| Name | | Location | Date DP Submitted | Date DP Approved | Cleanup Criteria | Projected Removal | Site Summ. Pg. No. |
|---|---|---|---|---|---|---|---|
| 11 | Homer Laughlin | Newell, WV | 1/95 | 1/95 | LTR-UNRES | TBD | Page C-18 |
| 12 | Jefferson Proving Ground | Madison, IN | 8/99 revised 6/02, 6/10* | 10/02 8/10* | LTR-RES | 9/10 | Page C-20 |
| 13 | Kaiser Aluminum | Tulsa, OK | (Phase 1) 8/98 (Phase 2) 5/01 | 2/00 6/03 | Action-UNRES LTR-UNRES | 10/06 | Page C-21 |
| 14 | Kerr-McGee | Cimarron, OK | 4/95 | 8/99 | Action-UNRES | 5/10 | Page C-22 |
| 15 | Mallinckrodt Chemical Inc. | St. Louis, MO | (Phase 1) 11/97 (Phase 2) 11/03 | 5/02 5/07* ** | LTR-UNRES | 7/08 | Page C-4 |
| 16 | Molycorp, Inc. – Washington | Wash., PA | 6/99 | 8/00 | Action-UNRES | 6/08 | Page C-26 |
| 17 | NWI Breckenridge | Breckenridge, MI | 3/04 | 8/04 | LTR-UNRES | TBD | Page C-28 |
| 18 | Pathfinder | Souix Falls, SD | 2/04 | 5/05 | LTR-UNRES | 5/07 | Page C-30 |
| 19 | Quehanna (formerly Permagrain Products, Inc.) | Media, PA | 4/98, revised 3/03, 3/06 | 7/98, 9/03, 11/06* | LTR-UNRES | 5/07 | Page C-31 |

15

**Table 2-3**
**Complex Decommissioning Sites**

| Name | | Location | Date DP Submitted | Date DP Approved | Cleanup Criteria | Projected Removal | Site Summ. Pg. No. |
|---|---|---|---|---|---|---|---|
| 20 | Royersford Wastewater Treatment Facility | Royersford, PA | TBD | TBD | LTR-UNRES | TBD | Page C-32 |
| 21 | Safety Light Corp. | Bloomsburg, PA | 12/00 | 12/01 | LTR-UNRES | 12/07 | Page C-34 |
| 22 | Salmon River | Salmon, ID | TBD | TBD | LTR-UNRES | 5/12 | Page C-36 |
| 23 | S.C. Holdings | Kawkawlin, MI | 11/03 | 3/06 | LTR-UNRES | 11/06 | Page C-37 |
| 24 | Shieldalloy Metallurgical Corp. | Newfield, NJ | 6/06 | 9/07* | LTR-RES | 9/10 | Page C-38 |
| 25 | Stepan Chemical Company | Maywood, NJ | NA | NA | LTR-UNRES | 9/09 | Page C-40 |
| 26 | Superbolt (formerly Superior Steel) | Pittsburgh, PA | TBD | TBD | LTR-UNRES | TBD | Page C-42 |
| 27 | UNC Naval Products | New Haven, CT | 8/98 revised 2004 | 4/99 11/06* | LTR-UNRES | 10/07 | Page C-43 |
| 28 | West Valley | West Valley, NY | 2007* | 2008* | LTR-UNRES*** | TBD | Page C-44 |
| 29 | Westinghouse Electric (Churchill Facility) | Pittsburgh, PA | 5/05 | NA+ | LTR-UNRES | TBD | Page C-46 |
| 30 | Westinghouse Electric (Hematite Facility) | Jefferson City, MO | 4/04 revised 6/06 | 9/07* | LTR-UNRES | 3/10 | Page C-47 |

## Table 2-3
## Complex Decommissioning Sites

| Name | | Location | Date DP Submitted | Date DP Approved | Cleanup Criteria | Projected Removal | Site Summ. Pg. No. |
|---|---|---|---|---|---|---|---|
| 31 | Westinghouse Electric (Waltz Mill) | Madison, PA | 4/97 | 1/00 | LTR-UNRES | 10/07 | Page C-49 |
| 32 | Whittaker Corp. | Greenville, PA | 12/00 Revised 8/03, 10/06* | 5/07* | LTR-UNRES | 2/08 | Page C-50 |

\*    Estimated Date

\*\*   Timeline for approving DP is protracted because of: (a) satisfying National Environmental Policy Act (NEPA) requirements; (b) conduct of public hearing; (c) multi-phase DP submittals; or (d) combination of all the above.

\*\*\*  The West Valley DP has not yet been submitted. The staff anticipates that West Valley DP will include plans to release a large portion of the site for unrestricted use, and the remainder of the site may have a perpetual license or be released with restrictions.

+   The Westinghouse Electric (Churchill Facility) submitted a DP before notifying NRC that it had permanently ceased operations. The licensee intends to continue licensed operations for several more years. The staff reviewed the DP and will provide a summary of its findings regarding the adequacy of the DP, but will not approve it. When the licensee permanently ceases operations, a new DP will need to be submitted to NRC for review and approval.

NOTES:

- The cleanup criteria identified in this table present the staff's most recent information, but not necessarily represent the current or likely outcome.

- Abbreviations used in this table include Action for SDMP Action Plan Criteria, LTR for LTR Criteria, RES for Restricted Use, and UNRES for Unrestricted Use.

**Table 2-4**
**Decommissioning Title II Uranium Recovery Sites**

| | Name | Location | DP Approved | Completion of Decomm. | Site Summ. Pg. No. |
|---|---|---|---|---|---|
| 1 | American Nuclear Corporation | Casper, WY | 10/88, Revision 2006 | TBD | Page D-1 |
| 2 | Bear Creek | Converse County, WY | 5/89 | 2007 | Page D-2 |
| 3 | COGEMA Mining Inc. | Mills, WY | 12/01 | TBD | Page D-3 |
| 4 | ExxonMobil Highlands | Converse County, WY | 1990 | 2008 | Page D-4 |
| 5 | Homestake | Grants, NM | Revised plan - 3/95 | 2017 | Page D-5 |
| 6 | Pathfinder - Lucky MC | Gas Hills, WY | Revised plan - 7/98 | 2007 | Page D-6 |
| 7 | Pathfinder - Shirley Basin | Shirley Basin, WY | Revised plan - 12/97 | TBD | Page D-7 |
| 8 | Rio Algom - Ambrosia Lake | Grants, NM | 2003 (mill); 2004 (soil) | 2009 | Page D-8 |
| 9 | Sequoyah Fuels Corporation | Gore, OK | 2007 | TBD | Page D-9 |
| 10 | Umetco Minerals Corporation | East Gas Hills, WY | Revised soil plan - 4/01 | 2010 | Page D-10 |
| 11 | United Nuclear Corporation | Churchrock, NM | 3/91, Revision 2005 | TBD | Page D-11 |
| 12 | Western Nuclear Inc.- Split Rock | Jeffrey City, WY | 1997 | 2008 | Page D-13 |

## 2.4.1 Uranium Recovery Facility Decommissioning Process

The decommissioning process is initiated by any one of the following conditions:

1. The license expires or the license is revoked;
2. The licensee has decided to permanently cease principal activites at the entire site or in any separate building or outdoor area;
3. No principal activities have been conducted for a period of 24 months; or
4. No principal activities have been conducted for a period of 24 months in any separate building or outdoor area.

Several major steps comprise the decommissioning process including the following: (a) notification; (b) submittal and review of the reclamation plan (RP); (c) implementation of the RP; (d) completion of reclamation; (e) construction-completion report review and inspection; (f) well-field restoration report review; (g) license termination; and (h) transfer of property to the long-term care custodian.

### Notification

Within 60 days of the occurrence of any of the triggering conditions, the licensee must notify NRC of such occurrence and either begin decommissioning or, if required, submit an RP within 12 months of notification and begin decommissioning upon plan approval. Two exceptions to this exist. First, for new ISL or conventional facilities, groundwater restoration, surface reclamation, and facility DPs are submitted with the initial license application. These plans are reviewed and approved before a license is issued. For ISLs, reclamation can occur at one well field, while others are being actively mined. Second, under 10 CFR 40.42(f) (timeliness in decommissioning requirements), facilities may delay decommissioning if NRC determines that such a delay is not detrimental to public health and the environment and it is in the public interest. Such a delay has been granted on multiple occasions to one conventional mill facility in standby status.

### RP - Existing Facilities

At this point in time, all uranium recovery facilities in existence before the enactment of the Uranium Mill Tailings Radiation Control Act of 1978 have NRC-approved RPs. Therefore, staff would only review amendments to existing RPs for such existing facilities. Amendments to RPs would be required under the following circumstances:

- Environmental contamination or other conditions not considered in the existing RP;

- A change in reclamation procedures that the licensee had requested.

Depending on the complexity of the revision, a meeting between the licensee and NRC staff may be warranted. This meeting would serve to make the RP amendment process more efficient by reducing the need for multiple RAIs.

After the amended RP is submitted, the review process begins with an acceptance review. Acceptance reviews are generally administrative in nature and include, but are not limited to, the following: (a) completeness of the application; (b) legibility of drawings; (c) adequacy of information; (d) justification for proprietary information; and (e) obvious technical inadequacies. Acceptance reviews are used to verify that the application contains sufficient information before the staff begins detailed technical reviews. Furthermore, a limited technical review would be conducted.

Amendments to RPs would require either an EA or an EIS, depending on amendment complexity. If staff determine that an EIS is required, staff will take the following basic EIS development steps:

- Notice of Intent;
- Public scoping meeting;
- Preparation and publication of the scoping report;
- Preparation and publication of the draft EIS;
- Public comment period on the draft EIS, including a public meeting;
- Preparation and publication of the final EIS; and
- Preparation and publication of the ROD.

If an EA is adequate for the RP amendment, staff will issue a draft EA to cooperating agencies, incorporate agency comments, and publish the final EA and a FONSI. Staff will also prepare the technical evaluation report (TER) concurrently with EA or EIS preparation. After publication of the FONSI or ROD, a license amendment and TER will be issued, approving the RP amendment, along with any additional license conditions deemed necessary from the environmental and technical review processes.

Regardless of whether an EA or EIS is developed, the staff structures its reviews to minimize the quantity of RAIs without diminishing the technical quality or completeness of the licensee's ultimate submittal. For example, staff will develop a set of needs and clarifications and discuss those needs to determine the necessity of all the requests before addressing the licensee. Staff will subsequently discuss those needs and clarifications, with the licensee, before developing the final RAI, to provide the licensee with an opportunity to address the straightforward needs or clarifications immediately.

RP - New Facilities

Procedures for reviewing RPs for new facilities are similar to those for existing facilities. However, because new facility RPs are incorporated into the license application, an EIS is automatically required for the RP, per 10 CFR 51.20(8).

Implementation of the RP

After approval of the RP, the licensee must complete decommissioning, in accordance with the approved RP, within 24 months, or apply for an alternate schedule. For conventional facilities, with groundwater contamination or ISL uranium extraction facilities, 24 months are usually insufficient to complete groundwater reclamation, because groundwater contamination is more difficult to remediate than surface contamination. NRC staff will inspect the licensee, during decommissioning operations, to ensure compliance with the RP, license conditions, and NRC, and other applicable regulations (i.e., US Department of Transportation regulations).

## Completion of Decommissioning

Decommissioning involves two different activities, surface reclamation (i.e., surface contamination, 11e.(2) byproduct material, and structures) and groundwater reclamation. Groundwater reclamation is considered completed when concentrations on and offsite (depending on the extent of contaminant migration) meet previously established groundwater protection standards (GPS') per Part 40, Appendix A. Three types of standards have been established per Criterion 5B(5), in Appendix A. These are as follows:

- Commission-approved background concentrations;

- Representative values presented in Table 5C in Appendix A; and

- Alternate concentration limits (ACLs).

When GPS' are originally established, the values are generally Commission-approved background or values, Table 5C, or EPA maximum concentration levels (MCLs), per the Safe Drinking Water Act. If the licensee demonstrates that concentrations of certain constituents cannot be restored to either background or MCLs, then the staff may approve alternate concentration limits (ACLs), after considering all the items found in Appendix A, Criterion 5B(6).

To obtain ACLs, the licensee submits a license amendment application and a detailed environmental report that addresses all the items in Criterion 5B(6). If the staff determines that the ACLs are protective of public health and the environment, the staff may approve the ACLs. Staff documents its review by publishing an EA and FONSI, and issuing a TER. After ACLs are approved, groundwater reclamation may cease and surface reclamation may be completed. However, ACL amendments will incorporate groundwater and surface water (if needed) monitoring programs that continue after reclamation is finished.

After surface reclamation is completed, the licensee issues a construction completion report for staff review and approval. As part of this review, staff performs a construction completion inspection to confirm that surface reclamation was performed according to the RP, license conditions, and NRC regulations. Inspections also include surveys of tailings disposal areas, to ensure that radon emissions comply with Part 40, Appendix A, Criterion 6. If additional information is required, staff will issue an RAI to address outstanding issues. After all issues are resolved, staff will publish an EA and FONSI. Staff will subsequently issue a license amendment and TER documenting the staff's review and approving the construction completion report.

## License Termination - Conventional Mills

After all reclamation activities have been completed and approved, the licensee, NRC staff, and the long-term custodian will start license termination procedures. Before a conventional mill license is terminated, the custodial agency (i.e., State agency, DOE, or other Federal agency) will submit a long-term surveillance plan (LTSP) for staff review and concurrence. The LTSP documents the custodian's responsibilities for long-term care, including security, inspections, groundwater and surface water monitoring, and remedial actions. After staff approves an

LTSP, the license may be terminated and title to the site, including all disposal areas, is transferred to the custodian. After a license is terminated, the custodian is regulated under a 10 CFR 40.28 general license.

In some cases, groundwater contamination has migrated offsite and cannot be reclaimed. In these cases, the licensee must purchase the offsite properties or otherwise establish institutional controls (ICs) over the land and groundwater use to prevent human exposures to contaminated groundwater. All land, beyond the site boundary, that is purchased or otherwise regulated by ICs, is incorporated into the long-term surveillance boundary. Use of ICs represents an alternative to the regulations in Part 40, Appendix A. However, staff may consider the use of ICs, if the licensee can demonstrate that it is protective of public health and the environment.

License Termination - ISL Uranium Extraction facility

License termination at an ISL uranium extraction facility occurs when all groundwater and surface reclamation is completed. After reclamation, well-field restoration and surface restoration reports are reviewed and approved by the staff. Surface restoration reports would typically include an inspection. Because ISL uranium extraction facility owners are prohibited by regulation to dispose of 11e.(2) byproduct material at their sites, long-term care is not required. Thus all groundwater and surface reclamation is performed for unrestricted release, and all land occupied by the ISL facility is returned to the original owner.

2.4.2    Summary of FY 2006 Activities

In FY 2006, Uranium Recovery staff completed approximately 30 licensing actions. The most significant of the decommissioning actions are as follows:

- Approval of the ACL amendment for the Pathfinder Mines Corporation, Shirley Basin site;

- Approval of the ACL amendment for the Rio Algom Mining Corporation, Ambrosia Lake site;

- Approval of the ACL amendment for the Western Nuclear, Inc., Split Rock site;

- Approval of the COGEMA - Irigaray Mine restoration report; and

- Approval of revisions to the GPS' at the United Nuclear Corporations, Church Rock site.

2.5    Fuel Cycle Facility Decommissioning

FCSS regulates facilities that enrich uranium and fabricate it into fuel for use in nuclear reactors, and facilities that fabricate nuclear fuel that is a combination of uranium and plutonium oxides. Several types of fuel cycle facilities are licensed for the enrichment and fabrication of uranium into nuclear fuel used for nuclear power plants. These include uranium fuel fabrication facilities, uranium hexafluoride production (conversion) facility, and gaseous diffusion enrichment facilities. Most of these facilities have been in operation for 20 or more years. As technology improves and operations at these facilities change, there are often unused areas on the sites that have residual contamination. The NRC staff continues to work closely with the States and EPA to regulate remediation of unused portions of fuel cycle facilities.

Regulation of fuel cycle facilities is accomplished through a combination of: (a) regulatory requirements; (b) licensing; (c) safety oversight, including inspection, assessment of performance, and enforcement; (d) operational experience evaluation; and (e) regulatory support activities. Table 2-5 identifies the fuel cycle facilities undergoing decommissioning. Facility status summaries are provided in Appendix E.

## Table 2-5
## Fuel Cycle Facilities Undergoing Decommissioning

| | Name | Location | Status | Site Summ. Pg. No. |
|---|---|---|---|---|
| 1 | AREVA NP | Richland, WA | Active | Page E-1 |
| 2 | General Atomics | San Diego, CA | Active | Page E-2 |
| 3 | Honeywell | Metropolis, IL | Active | Page E-3 |

### 2.5.1 Fuel Cycle Facility Decommissioning Process

In general, the decommissioning process for fuel cycle facilities and complex material sites is the same (see Section 2.3.1). Decommissioning activities at fuel cycle facilities can be conducted during operations (partial decommissioning) or after the licensee has ceased all operational activities.

Project management responsibility for fuel cycle facilities resides in FCSS during licensee operations, and within DWMEP during entire site decommissioning in support of license termination. Project management responsibility for fuel cycle facilities is transferred from FCSS to DWMEP when: (1) the licensee has ceased all operational activities; and (2) a critical mass of material no longer remains at the site.

### 2.5.2 Summary of FY 2006 Activities

In 2006, one conversion facility (Honeywell) and two fuel manufacturers (AREVA NP and General Atomics) continued some decommissioning activities.

### 3. Guidance and Rulemaking Activities

In FY 2006, the staff completed a number of guidance and rulemaking activities. These activities resulted from the Integrated Decommissioning Improvement Plan (IDIP), Rev. 1, published in March 2005. The IDIP described how the staff planned to implement recommendations from the Decommissioning Program Evaluation, the LTR Analysis recommendations approved by the Commission, Commission direction resulting from the 2004 annual decommissioning briefing, and other improvements.

The 2005 annual decommissioning report identified a number of follow-up actions the staff intended to take to implement the IDIP in FY 2006. Updates to the IDIP, based on staff assessments, staff decommissioning experience, and independent program reviews, such as the Office of the Inspector General (OIG) audits result in "continuous improvement" of the Decommissioning Program.

Major IDIP improvement activities completed in FY 2006 include:

- Finalized guidance in NUREG-1757, Supplement 1, "Consolidated NMSS Decommissioning Guidance: Updates to Implement the License Termination Rule Analysis," which included guidance on issues associated with implementing the LTR in Part 20, Subpart E, including: (a) restricted use; (b) onsite disposal; (c) realistic scenarios; (d) removal of material after license termination; (e) engineered barriers; and (f) intentional mixing of soil;

- Began development of a proposed rule and supporting guidance for preventing future legacy sites (i.e., sites with inadequate funding to complete decommissioning). These actions will eventually resolve the LTR Analysis issues regarding financial assurance and facility operational releases that have resulted in decommissioning difficulties.

- Continuation of improvements to collect, document, and disseminate decommissioning lessons-learned, including: (a) updating the decommissioning web page for lessons learned; and (b) exchanging information on lessons-learned with stakeholders at the March 2006 meeting with the Fuel Cycle Facility Forum, OAS, and Nuclear Energy Institute;

- Published DWMEP Operations Manual to put in place new procedures that implement program improvements including: (a) staff expenditure tracking; (b) prioritization of work; (c) operating plan management; (d) planning for revised guidance; (e) sharing information; (f) updating the IDIP; (g) independent reviews; and, (h) defining the roles of the offices and divisions involved with the Comprehensive Decommissioning Program;

## 4. Research Activities

The Office of Nuclear Regulatory Research (RES) continued providing information to NMSS to support dose modeling of releases of radioactive material from decommissioning sites. In addition to research activities, RES staff provided technical support to NMSS for Cimarron and West Valley, and developed input for the final version of revised decommissioning guidance on the use of engineered barriers.

RES is continuing the development or modification of a number of computer codes useful for site decommissioning analyses, including: (a) modifying dose-assessment codes, to incorporate added realism; (b) bench-marking RESRAD-OFFSITE to compare its capabilities to those of other commonly used dose codes; (c) developing FRAMES2 (Framework for Risk Assessment of Multimedia Environmental Systems) with a linkage to the Department of Defense Groundwater Modeling System, and training NMSS staff in the use of the linked codes; and (d) providing NMSS with a report on new conceptual models for food-chain pathways . A new contract to support further development of Spatial Analysis and Decision Assistance was placed to provide tools for more efficiently designing site characterization of contaminated sites, assessing risk, determining the location of future samples, and designing remedial action. During the past year, RES also provided training to NMSS, NRR, and Regional staff on the assessment of uncertainty in groundwater modeling and the design of monitoring systems to assess groundwater contamination.

RES completed work on modeling the fundamental processes controlling sorption reactions. It continued work on the practical application of reactive transport models, in performance assessments of chemically complex sites, and resolution of comments on methods for establishing financial assurance requirements for the decommissioning of in-situ leach mines. It also provided information on formal model abstraction techniques for selecting the right level of abstraction for a given degree of site or process complexity. Additionally, RES also continued work to further the understanding of the evolution and degradation of clay covers, through laboratory testing.

RES maintains two technical advisory groups (TAGs) that enhance communication on issues important to site decommissioning and provide feedback to RES on research direction. The TAGs are the "Technical Advisory Group on Groundwater and Performance Monitoring," and the "Technical Advisory Group on Assessing Uncertainty in Simulation Modeling of Environmental Systems." The TAG on ground-water issues was particularly useful this year in providing insights about the environmental contamination found at several operating nuclear power plants.

During the past year, RES staff also continued to support interagency cooperative activities. The RES staff, along with NMSS staff, continued to participate in activities of the ISCORS, and RES staff supported the Interagency Steering Committee on Multimedia Environmental Models (ISCMEM). During this year, the memorandum of understanding (MOU) that created ISCMEM was renewed by seven participating federal agencies (for details see www.ISCMEM.org).

5.    **International Activities**

The DWMEP interacts with international organizations and governments in a number of ways including: (a) the International Atomic Energy Agency (IAEA); (b) the Organization for Economic Cooperation and Development's Nuclear Energy Agency; (c) bilateral and trilateral exchanges with other countries; (d) hosting foreign assignees and providing reciprocal assignments; (e) developing and providing workshops to requesting countries; and (f) providing technical support as needed to the NRC Office of International Programs. NRC generally is recognized in the international nuclear community as an experienced leader in the decommissioning of nuclear sites. NRC staff interaction with international organizations and governments allows NRC to share insights into decommissioning approaches that are successful, safe, and cost-effective. It also allows the NRC staff to provide input into the various international guidance and requirements that NRC and NRC licensees will need to consider as they interact in a global environment. The NRC staff gains insight into approaches and methodologies that are being employed in the international community and considers these approaches as they continue to risk-inform the NRC Decommissioning Program. A summary of the most significant of these activities is provided below.

<u>IAEA Activities</u>

The NRC decommissioning staff participated in the development of the IAEA Safety Standards Series. Within the past year, staff supported the IAEA by:

- Participating in the Joint Convention on the Safety of Spent Fuel Management and on the Safety of Radioactive Waste Management, by completing Joint Convention country report reviews.

- Participating in twice-yearly meetings of the IAEA Waste Safety Standards Committee, which address decommissioning specifically, as part of the IAEA's waste safety activities.

- In March 2006, staff traveled to Kiev, Ukraine, for expert assistance in the preparation of a draft DP for Units 1-3 of the Chernobyl Nuclear Power Plant.

<u>Bilateral and Trilateral Exchanges with Other Countries</u>

A delegation from Spain visited NRC in FY 2006 to discuss many topics associated with radioactive waste management. Facility decommissioning, especially for nuclear power plants, is usually of significant interest to the visiting delegations.

Staff also participated in a meeting with a Ukrainian Delegation in April 2006 – DWMEP topics included U.S. regulations and experience in specific regulatory areas, such as legal bases for establishing, and functioning of, decommissioning funds and funds for radwaste management.

Staff met with a delegation from the Korean Institute of Nuclear Safety, in July 2006, to discuss the decommissioning of nuclear reactors.

<u>Nuclear Energy Agency Activities</u>

- Annual Meeting / Topical Session; and

- Updating National Fact Sheets.


6. **Program Integration**

The staff continues to take steps to ensure integration of decommissioning activities. First, NMSS, NRR, RES, and Regions mutually track and coordinate decommissioning activities. Second, the Decommissioning Management Board meets bi-monthly to provide management input on decommissioning activities and issues. The Board, composed of managers from NMSS, RES, NRR, and the Regions, along with the Office of the General Counsel, serves as an effective mechanism for integrating interoffice and interregional program activities and issue resolution. The Board is a mechanism by which the staff has enhanced intraagency communication, and it ensures that NRC's regulatory processes are integrated. Third, Headquarters and Regional staff held a Decommissioning Counterparts Meeting, in May 2006, to discuss issues affecting the decommissioning program. In FY 2006, consolidation of the decommissioning program began with consideration of re transfer or RTRs from NRR to NMSS.

As noted in Section 2.1, the transition occurred on October 1, 2006. Finally, RES, NRR, the regions, and Agreement States participate on review teams to comment on draft decommissioning guidance.

## 7.    Agreement State Activities

As stated in Section 2 of this report, 34 States have signed formal agreements with NRC, by which those States have assumed regulatory responsibility over certain byproduct, source, and small quantities of special nuclear material, including decommissioning of some complex materials sites. However, once a State becomes an Agreement State, NRC continues to have formal and informal interactions with the State.

Formal interactions with Agreement States in FY 2006 include:

- IMPEP reviews of 11 Agreement States;

- Organization of Agreement States (OAS) participation on the Division of Industrial and Medical Nuclear Safety (IMNS) working group to develop the proposed rule to prevent future legacy sites;

- Agreement State representative participation on the writing and review teams revising NUREG-1757; and

- DWMEP staff participation in Conference of Radiation Control Program Directors E-24 Subcommittee on Decommissioning.

Examples of informal interactions include:

- OAS participation at the 2006 Annual Decommissioning Commission briefing;

- DWMEP staff attendance at the annual OAS meeting;

- DWMEP and Regions coordinated and interacted with States on specific decommissioning sites and issues (Yankee Rowe, Connecticut Yankee, Union Carbide, FMRI, Kerr McGee Cushing and Cimarron, Indian Point Unit 1, Heritage Minerals, Inc., Shieldalloy Metallurgical Corp., and the WVDP); and

- DWMEP and Regional coordination with Pennsylvania Department of Environmental Protection, in preparation for Pennsylvania becoming an Agreement State (quarterly conference calls to discuss the status of decommissioning activities at complex sites and Pennsylvania observation of NRC inspections).

Table 7-1 identifies the decommissioning and uranium recovery sites in the Agreement States. In FY 2007, NRC staff will work with the Agreement states to incorporate information on Decommissioning activities in Agreement states into the annual report.

**Table 7-1**

**Decommissioning and Uranium Recovery Sites in Agreement States**

| State | Name | Location | Date DP Submitted | Date DP Approved | Clean-up Criteria | Project Complete |
|-------|------|----------|-------------------|------------------|-------------------|------------------|
| AL | | | No Sites | | | |
| AZ | | | No Sites | | | |
| AR | Harmon Road LLRW Disposal Site (Arkansas University) | Fayetteville, AR | | | | TBD |
| AR | SEFOR (Research Reactor at University of Arkansas) | Fayetteville, AR | | | | TBD |
| CA | General Atomics | San Diego, CA | 10/14/96 | 8/26/97 | Surface- & concentration-based criteria | 12/07 |
| CA | ICN Biomedicals | Irvine, CA | 11/14/05 | 5/15/06 | Surface- & concentration-based criteria | 3/07 |
| CA | Excel Research Services, Inc | Fresno, CA | 5/03 | TBD | Concentration-based criteria | TBD |
| CA | Providencia Holdings, Inc. | Burbank, CA | 7/16/01 | 10/31/02 | Surface- & concentration-based criteria | 12/06 |
| CA | Molycorp, Inc. – Mountain Pass Plant | Mountain Pass, CA | 6/9/06 | | Concentration-based criteria | 3/07 |
| CA | Aerojet Ordnance Company | Chino Hills, CA | 2/15/96 | 5/31/96 | Surface- & concentration-based criteria | 12/07 |

28

**Table 7-1**

**Decommissioning and Uranium Recovery Sites in Agreement States**

| State | Name | Location | Date DP Submitted | Date DP Approved | Clean-up Criteria | Project Complete |
|-------|------|----------|-------------------|------------------|-------------------|------------------|
| CA | PTRL West Inc. | Hercules, CA | 2/7/00 | 4/6/00 | Indistinguishable from background | 12/07 |
| CA | Kirk Rich Dial Company | Los Angeles, CA | N/A | N/A | Indistinguishable from background | TBD |
| CA | ABC Management Inc./ dba ABC Laboratories | Madera, CA | 4/7/93 | 12/21/94 | Surface- & concentration-based criteria | TBD |
| CA | MP Biomedicals | Irvine, CA | 5/17/06 | TBD | Surface-based criteria | TBD |
| CO | Umetco | Uravan, CO | | 02/01/1987 | Criterion 6(6) | 2008 |
| CO | Umetco Maybell | Maybell, CO | 01/01/1995 | 1995 | Criterion 6(6) | 2005 |
| CO | Cotter Uranium Mill | Canon City, CO | Revised 2005 | 2005 | Criteria 6(6) - restricted area for soils. Surface- & concentration-based - some Superfund units and licensed portion | In standby. TBD if going into D&D |
| CO | Schwardzwalder Mine (Cotter) | Golden, CO | 12/01/1996 | 1997 | Criterion 6(6) | TBD |
| CO | Hecla Durita | Naturita, CO | 1991 | 1993 | Criterion 6(6) | Completed 1998 |
| CO | CSMRI Table Mtn. | Golden, CO | 08/01/2006 | TBD | Criterion 6(6) | 2007 |
| CO | CSMRI Creekside | Golden, CO | TBD | TBD | TBD | 2007 |

**Table 7-1**

**Decommissioning and Uranium Recovery Sites in Agreement States**

| State | Name | Location | Date DP Submitted | Date DP Approved | Clean-up Criteria | Project Complete |
|-------|------|----------|-------------------|------------------|-------------------|------------------|
| CO | Sweeney Mining and Milling | Boulder, CO | Pending | | TBD | TBD |
| CO | Homestake Mining and Pitch | Sargeants, CO | 05/01/2001 | 06/01/2001 | Criterion 6(6) | TBD |
| CO | Redhill Forest | Fairplay, CO | Pending | TBD | 25 mrem | TBD |
| CO | Clean Harbors | Deer Trail, CO | 2005 | 2006 | 25 mrem | TBD |
| CO | Cyprus Amax | Golden, CO | 01/01/2005 | 05/01/2005 | 5 pCi/g Ra-226 > bkgd. | 2006 |
| FL | Mosaic Fertilizer, LLC | Nichols, FL | 6/3/05 | Pending | <25mrem/yr | 2007 |
| FL | U.S. Agri-Chemicals Corp. | Fort Meade, FL | 3/13/06 | Pending | <25mrem/yr | None |
| FL | C.F. Industries, Inc. | Bartow, FL | Pending | N/A | N/A | N/A |
| FL | Piney Point Phosphates, Inc. | Bradenton, FL | Pending | TBD | TBD | TBD |
| GA | | No Sites | | | | |
| IA | | No Sites | | | | |
| IL | Chicago Magnesium | Blue Island, IL | 11/02/02 | 02/01/04 | Surface- & concentration-based criteria | Ph 1-12/04 Ph 2- 8/06 Ph 3- unk |

**Table 7-1**

**Decommissioning and Uranium Recovery Sites in Agreement States**

| State | Name | Location | Date DP Submitted | Date DP Approved | Clean-up Criteria | Project Complete |
|---|---|---|---|---|---|---|
| IL | Spectrulite | Madison, IL | 01/01/05 | 06/01/05 | Surface- & concentration-based criteria | 11/06/06 |
| IL | Tronox (Kerr-McGee) (Uranium Recovery Site) | West Chicago, IL | 09/01/93 | 09/01/94 | Concentration-based criteria | Completed 11/05 |
| | | | | | Groundwater: Part 332.230 references 10 CFR Part 40, Appendix A | Unknown |
| KS | | No Sites | | | | |
| KY | State did not reply to information request | | | | | |
| LA | | No Sites | | | | |
| ME | | No Sites | | | | |
| MD | | No Sites | | | | |
| MA | Shpack Landfill | Norton, MA | 09/04 | 09/04 | <10 mrem/yr | |
| MA | Yankee Rowe Nuclear Power Plant | Rowe, MA | 12/05 | 12/05 | < 10 mrem/yr < 20,000 pCi/L H-3(GW) | 6/07 |
| MN | | No Sites | | | | |
| MS | | No Sites | | | | |

**Table 7-1**

**Decommissioning and Uranium Recovery Sites in Agreement States**

| State | Name | Location | Date DP Submitted | Date DP Approved | Clean-up Criteria | Project Complete |
|---|---|---|---|---|---|---|
| NE | | | No Sites | | | |
| NV | | | No Sites | | | |
| NH | Kollsman, Inc. | Merrimac, NH | 5/05 | 2/06 | <10 mrem/year | 12/06 |
| NM | | | No Sites | | | |
| NY | | | No Sites | | | |
| NC | | | No Sites | | | |
| ND | | | No Sites | | | |
| OH | RMI Environmental Services, Inc. | Ashtabula, OH | 4/27/95 | 9/97 | Surface- & concentration-based criteria | 12/31/06 |
| OH | Metallurg Vanadium Corp. (Formerly Shieldalloy Metallurgical Corp.) | Cambridge, OH | 7/13/99 | 3/6/02 | Concentration-based criteria | 7/31/07 |
| OH | Ineos USA, LLC. (Formerly BP Chemicals, Inc.) | Lima, OH | 4/16/92 | 6/98 | Concentration-based criteria | 8/18/03 |
| OH | Advanced Medical Systems, Inc. | Cleveland, OH | 5/25/04 | 7/5/05 | Surface- & concentration-based criteria | |
| OK | | | No Sites | | | |
| OR | TDY Industries dba Wah Chang | Albany, OR | 6/11/03 | 3/08/06 | <25 mrem/yr | TBD |

**Table 7-1**

**Decommissioning and Uranium Recovery Sites in Agreement States**

| State | Name | Location | Date DP Submitted | Date DP Approved | Clean-up Criteria | Project Complete |
|---|---|---|---|---|---|---|
| OR | PCC Structurals, Inc. | Portland, OR | 6/10/06 | 9/14/06 | <25 mrem/yr | TBD |
| RI | | | No Sites | | | |
| SC | | | No Sites | | | |
| TN | | | No Sites | | | |
| TX | ExxonMobil (Uranium Recovery Site) | Live Oak Co., TX | 4/85 | 9/82 | Concentration-based criteria | TBD |
| TX | ConocoPhillips (Uranium Recovery Site) | Karnes Co., TX | 11/87 | 9/80 | Concentration-based criteria | TBD |
| TX | Rio Grande Resources (Uranium Recovery Site) | Karnes Co., TX | 4/93 ACL - 11/97 | 11/96 | Concentration-based criteria | TBD |
| TX | COGEMA (Uranium Recovery Site) | Duval Co., TX | 11/03 | 4/06 | Concentration-based criteria | GW complete Surface - ongoing |
| TX | Intercontinental Energy Corp. (Uranium Recovery Site) | Live Oak Co., TX | 3/03 | Ongoing | Concentration-based criteria | GW complete Surface - 12/07 |

33

**Table 7-1**

**Decommissioning and Uranium Recovery Sites in Agreement States**

| State | Name | Location | Date DP Submitted | Date DP Approved | Clean-up Criteria | Project Complete |
|---|---|---|---|---|---|---|
| TX | Everest Exploration, Inc. (decommissioning of Tex-1 Mt. Lucas sites) | Karnes and Live Oak Counties | 8/01 | Ongoing | Concentration based criteria | GW complete<br><br>Surface clean-up ongoing |
| UT | Rio Algom Uranium Mill | Lisbon Valley, UT | | | Concentration based criteria | TBD |
| WA | Dawn Mining Company (Uranium Recovery Site) | Ford, WA | 05/94 | 02/95 | Concentration based criteria | 12/13 |
| WI | | No Sites | | | | |

## 8. Resources

The total decommissioning program staff budget, for FY 2006 and FY 2007, is 96 full-time equivalents (FTEs) and 91 FTEs, respectively. These resource figures include: (a) licensing casework directly related to decommissioning sites; (b) inspections; (c) project management and technical support for decommissioning power reactors, uranium mill tailings facilities, and fuel cycle facilities; (d) development of rules and guidance; (e) environmental impact statements and EAs; (f) research to develop more realistic analytical tools to support licensing and rulemaking activities; and (g) Waste Incidental to Reprocessing. These figures also include supervisory and non-supervisory indirect FTEs associated with the decommissioning program.

## 9. FY 2007 Planned Programmatic Activities

A number of programmatic activities are planned for FY 2007. The most significant of these activities include: (1) consolidation of the Decommissioning Program; (2) continuation of IDIP improvement activities; (3) continuation of International activities; and (4) licensing of uranium mills.

As noted in Section 2.1, project management and oversight responsibility for several decommissioning reactors (Indian Point – Unit 1, Millstone – Unit 1, Vallecitos, and the Nuclear Ship Savannah) and 14 research and test reactors will transfer from NRR to FSME on October 1, 2006. The transfer will be accomplished in accordance with the "Transition Plan and Communication Plan for the Transfer of 14 Decommissioning Research and Test Reactors, Two Decommissioning Early Demonstration Reactors and Two Decommissioning Power Reactors."

Section 3 of this report identifies a number of IDIP improvement activities completed in FY 2006. In addition to the completed activities, the staff is making progress on a number of IDIP activities planned for completion in FY 2007. Major IDIP improvement activities planned for FY 2007 include:

- Preparing for, and participating in, an Office of Management and Budget (OMB) Program Assessment Rating Tool (PART) review, including a reevaluation of the decommissioning program and effectiveness of improvements. At OMB's request, the PART review was postponed from 2006;

- Continue with rulemaking concening groundwater restoration at ISLs;

- Publishing a proposed rule and draft guidance, for public comment, in 2007, for the rulemaking and supporting guidance on measures to prevent future legacy sites (changes to financial assurance and licensee operations);

- Draft inspection and enforcement guidance to enhance monitoring, and reporting procedures to prevent future legacy sites; and

- Implementing restricted release options in the license termination role at Shieldalloy and AAR sites.

In FY 2007, the staff will continue its interactions with IAEA and participation in bilateral and trilateral exchanges with other countries. One new activity of note for FY 2007, and possibly FY 2008, will be NRC's support of the IAEA technical assistance efforts to help the Iraqi Radioactive Source Regulatory Authority locate, secure, and regulate radioactive materials, and to decommission and manage the waste of the former Iraqi nuclear facilities.

As of September 2006, there is a renewed interest in uranium mining and milling. Consequently, approximately 10 entities have contacted the staff expressing interest in either submitting license applications for new mills (both ISL and conventional) or for satellite operations of existing mills. Furthermore, owners of two decommissioning sites and one conventional mill on standby have expressed interest in restarting operations. As a result, the staff expects to receive two applications to restart operations, in addition to at least two applications for new facilities and one application for a satellite facility, in FY 2007.

# Appendix A

# Site Summaries for
# Decommissioning Power Reactors

# Big Rock Point

## 1.0 Site Identification

Location:        Charlevoix, MI
License No.:     DPR-6
Docket No.:     50-155
License Status:  DECON
Project Manager:  Jim Shepherd

## 2.0 Site Status Summary

Big Rock Point (BRP) is located in Charlevoix County, Michigan, approximately four miles northeast of Charlevoix, Michigan, and approximately 11 miles west of Petoskey, Michigan, on the northern shore of Michigan's lower peninsula. The BRP site is owned by Consumers Energy Company (CE). The BRP Nuclear Plant was a boiling water reactor rated at 75MW electric, designed by General Electric Company.

The plant was permanently shut down on August 29, 1997. Fuel was transferred to the spent fuel pool by September 20, 1997. On September 19, 1997, CE submitted a post-shutdown decommissioning activities report (PSDAR) that identified decommissioning activities commencing in September 1997, and concluding in September 2002. The licensee selected the DECON option. On March 26, 1998, CE submitted a revised PSDAR that showed conclusion of decommissioning about August 2005. Dry fuel storage will continue through 2012 or later, depending on when the U.S. Department of Energy (DOE) accepts spent fuel.

CE submitted a license termination plan (LTP) on April 1, 2003. After negotiating a memorandum of agreement (MOA) with the Michigan State Historic Preservation Officer (SHPO), the U.S. Nuclear Regulatory Commission (NRC) approved it on March 12, 2005. CE is currently decommissioning the site for unrestricted use in accordance with the LTP. All systems and structures not needed for the independent spent fuel storage installation (ISFSI), except the intake piping and sanitary drainfield, have been removed. All remedial work has been completed and final status surveys will be completed by October, 2006.

All fuel was transferred to the ISFSI by March, 2003. Turbine building demolition and final status surveys (FSSs) were conducted during summer, 2005. The containment shell was removed in mid-2006. FSS reports and a request for partial site release are expected before the end of 2006.

After fuel is removed from the site to a DOE facility, the ISFSI will be decommissioned and the license terminated. CE plans for this to occur in 2012. DOE recently announced that it may start accepting commercial fuel at the Yucca Mountain site in 2017.

Contaminants at the site include uranium and decay products, and fission products. Low levels of ground water contamination, primarily tritium, are non-uniformly distributed at the site because of a dry, silty clay layer that underlies only the south part of the site. Boundaries between the geologic units are only approximated because of limited subsurface data. Reported radionuclide concentrations in ground water are generally less than the minimum detectable activity (MDA) except for tritium that is less than one half the U.S. Environmental

Protection Agency (EPA) drinking water standard of 20,000 pCi/l. Soil contamination is also generally below MDA.

## 3.0 Major Technical or Regulatory Issues

There is some public interest in the decommissioning of this site. The primary parties are the State of Michigan and the City Councils of surrounding areas. CE has an effective public outreach program and open communication with these parties.

The Michigan SHPO declared the facility itself eligible for the National Historic Register. Therefore, demolition is defined as an adverse effect that requires a MOA in accordance with 40 CFR 800. The MOA was negotiated to address the issues of early notification to SHPOs of NRC plans; documentation of the site using the Historic American Engineering Record System; and post-license termination access to the site by Native Americans, for whom the Big Rock is an historic gathering place. The MOA was executed by NRC, SHPO and CE in February, 2006.

## 4.0 Estimated Date For Closure

2007

# Dresden - Unit 1

## 1.0 Site Identification

Location:           Dresden, IL
License No.:        DPR-2
Docket No.:         50-0010
License Status:     SAFSTOR
Project Manager:    John Hickman

## 2.0 Site Status Summary

The plant shut down in October 1978 and is currently in SAFSTOR. The decommissioning plan (DP) was approved in September 1993. No significant dismantlement activities are underway. Asbestos removal, isolation of Unit 1 and Units 2 and 3, and general radiation cleanup activities are complete or in progress. The licensee will dismantle Unit 1 along with the other two units onsite, which is expected no earlier than 2011. The licensee submitted an updated PSDAR on June 1, 1998. The PSDAR public meeting was held on July 23, 1998.

Dresden Unit 1 produced power commercially from 1960 to October 1978, generating approximately 15,800,000 Megawatt-hours of electricity. The licensed power of the Unit was increased from 630 MWt to 700 MWt in September of 1962. The unit had a history of minor steam leaks and erosion in steam piping in the early and mid-1960s. There were also fuel failures during the period of September through December of 1964 and other times which, although not leading to off-gas releases above limits, did cause redistribution of radionuclides from the fuel to other parts of the primary system. Several systems in the plant used admiralty brass (Cu-Ni) heat exchange surfaces, including the Main Condenser. Most of these were taken out of service and replaced with stainless steel tubing. In the sixth partial refueling, the condenser was re-tubed from admiralty brass to 304L stainless steel. The use of Cu-Ni surfaces did lead to translocation and deposition of corrosion products throughout the operating systems. The use of carbon steel in the Secondary Feedwater System may have also contributed to the elevated corrosion radionuclide levels. These foregoing events led to the need to perform a chemical decontamination of the Primary System. The Unit was taken off-line on October 31, 1978, to backfit it with equipment to meet new federal regulations and to perform a chemical decontamination of major piping systems. While it was out of service for retrofitting, additional regulations were issued as a result of the March 1979 incident at Three Mile Island. The estimated cost to bring Dresden Unit 1 into compliance with these regulations was more than $300 million. Commonwealth Edison concluded that the age of the unit and its relatively small size did not warrant the added investment. In 1984, chemical decontamination of the primary system was performed and 753 curies of Cobalt-60 and 12.4 curies of Cesium-137 were removed. This decontamination was completed and activities began shortly thereafter to prepare the facility for decommissioning. In July of 1986, the NRC revised the Dresden Unit 1 license to possess-but-not-operate status. NRC approved Revision 3 to the Dresden Unit 1 DP on September 3, 1993. Subsequent revisions to the DP were reviewed and approved based on criteria similar to the criteria of Section 50.59 of Title 10 of the Code of Federal Regulations (10 CFR 50.59). In 1998, the DP was revised to the current Defueled

Safety Analysis Report (DSAR) format. The NRC approved the transfer of the facility licenses from Commonwealth Edison (ComEd) Company to Exelon Generation Company, LLC (Exelon) on January 12, 2001.

## 3.0 Major Technical or Regulatory Issues

The licensee is using the Holtec HISTAR 100 dual purpose cask and the HISTORM concrete overpack to store spent fuel. Casks have been loaded with Unit 1 spent fuel from the Unit 2 spent fuel pool, along with Unit 2 spent fuel, to address the Unit 2 spent fuel storage issue. In January 2002, the licensee completed transferring fuel from the Unit 1 spent fuel pool to dry storage.

## 4.0 Estimated Date For Closure

TBD

# Fermi

## 1.0 Site Identification

Location:           Newport, MI
License No.:        DRP-9
Docket No.:         50-16
License Status:     SAFSTOR
Project Manager:    Ted Smith

## 2.0 Site Status Summary

The licensee's initial stage of decommissioning is complete, and bulk sodium has been removed from the site. There is no spent fuel onsite and the facility is currently in SAFSTOR condition. The licensee is currently performing occupational safety enhancement activities; concentrating in non-radioactive areas, such as asbestos removal, and trace sodium cleanup. The trace sodium remediation effort is about 50 percent complete. The facility will be dismantled under the provisions of 10 CFR 50.59. The licensee plans to submit an LTP in 2007.

The Enrico Fermi Atomic Power Plant, Unit 1 (Fermi 1) was a fast breeder reactor power plant cooled by sodium and operated at essentially atmospheric pressure. The reactor plant was designed for a maximum capacity of 430 Mwt; however, the maximum reactor power with the first core loading (Core A) was 200 Mwt. The primary system was filled with sodium in December of 1960 and criticality was achieved in August 1963. Power ascension testing above 1 Mwt commenced in December 1965, immediately following receipt of the high power operating license. In November 1972, the Power Reactor Development Company made the decision to decommission Fermi 1. The fuel and blanket subassemblies were shipped offsite in 1973. The non-radioactive secondary sodium system was drained and the sodium sent to Fike Chemical Company. The radioactive primary sodium was stored in storage tanks and in 55 gallon drums until the sodium was shipped offsite in 1984. Decommissioning of the Fermi 1 plant was originally completed in December 1975. The site has been in a SAFSTOR status, awaiting final decommissioning. The license for Fermi 1 expires in 2025.

Current decommissioning cost estimate is $28-31 million (1998 dollars). Current amount in trust fund is $32 million.

## 3.0 Major Technical or Regulatory Issues

None.

## 4.0 Estimated Date For Closure

05/01/2008

# Haddam Neck - Connecticut Yankee

## 1.0 Site Identification

Location: East Hampton, CT
License No.: DPR-61
Docket No.: 50-213
License Status: DECON
Project Manager: Ted Smith

## 2.0 Site Status Summary

Steam generators, reactor coolant pumps, the pressurizer, the reactor vessel, and shield wall blocks from the Reactor Building shielding have been disposed offsite. All 1016 spent fuel assemblies and 2 casks of greater than Class C (GTCC) waste are stored at the ISFSI. The administration, primary auxiliary, and turbine buildings have been demolished. Removal of groundwater affecting soil near the former tank farm and containment building is complete. The containment structure and spent fuel building have been demolished, and only the radwaste reduction facility remains. The licensee's current plans will remove all structures from the site down to four feet below grade. This will leave concrete structures in only a few areas of the site, such as the reactor building basement, the discharge canal tunnel and part of the spent fuel building.

## 3.0 Major Technical or Regulatory Issues

None.

## 4.0 Estimated Date For Closure

2007

# Humboldt Bay

## 1.0 Site Identification

Location:          Eureka, CA
License No.:       DPR-7
Docket No.:        50-133
License Status:    SAFSTOR
Project Manager:   John Hickman

## 2.0 Site Status Summary

The plant was shut down in July 1976. A DP was approved in July 1988. Subsequent to the 1996 decommissioning rule, the licensee converted the DP into its DSAR which is now updated every two years. A PSDAR was issued by the licensee in February 1998. The plant is currently in SAFSTOR with incremental decommissioning activities ongoing. Decommissioning work at Humboldt Bay involves recently completed asbestos removal, currently in progress systems and structures radiological characterization, and future work on reactor and internals activation analysis, low-level waste (LLW) management plan development, developing of a work, cost, and scheduling process, and the developing of a facilities and staffing plan. This work phase will likely continue until a decision is made on accelerated decommissioning.

Humboldt Bay, Unit 3 is a 65 MWe boiling water reactor plant located 4 miles southwest of Eureka, CA. The plant operated commercially from 1963 to 1976 when it shut down for seismic modifications. In 1983, with the plant still shut down, PG&E determined that required seismic modifications and the requirements imposed as a result of the TMI-2 accident made continued operations no longer economically feasible, and therefore decided to decommission the plant. All fuel was placed in the spent fuel pool. A possess-but-not-operate license amendment was issued in 1985. In December of 2003, the licensee submitted and application to the NRC seeking regulatory approval for on-site dry cask storage of the fuel that is currently in the spent fuel pool. The licensee is evaluating a plan that would have all spent fuel in dry cask storage by 2008. Although the current license expires in 2015, the licensee is evaluating a plan that would have the plant dismantled, the Part 50 license terminated and site restoration completed in the 2009-2011 time frame. Humboldt Bay, Units 1 and 2 are fossil plants located immediately adjacent to Unit 3 and are still in commercial operation. The current decommissioning cost estimate is $333.6 million. There is currently $213.9 million in the decommissioning trust fund. The licensee estimates that an additional $41.5 million in revenue will be added to the decommissioning fund over the next 2.5 years. The remaining balance of $78.2 million to fully fund the decommissioning liability is presumed to accrue from fund investments, interest, and tax advantages during the next 10 years.

Currently, ISFSI licensing related activities are the major focus of the site. During the fall of 2003, the licensee began a detailed examination of the contents of its spent fuel pool in preparation for eventual removal of the fuel assemblies stored in the pool to a dry-cask-storage ISFSI. In the process of performing the spent fuel pool examination, the licensee discovered fuel rod fragments that could not be accounted for by records maintained at the facility. While in the process of performing a record review related to the fragment investigation, the licensee identified a discrepancy on June 23, 2004, that called into question the location of three segments of a portion of a single spent fuel rod removed from an assembly (designated A-49) in 1968. Records from 1968 indicate that a single fuel rod from assembly A-49 was cut into three 18-inch segments that were placed in a small container with an intention to ship them to an off-site lab for analysis. The records indicate that the off-site shipment never occurred and that the three 18-inch segments in their special storage container were placed somewhere in the spent fuel pool without identifying the specific location. The licensee has been unable to locate these three 18-inch rod segments in the spent fuel pool and has not found any records documenting their shipment off site. On May 27, 2005, PG&E issued their special nuclear material (SNM) Control and Accountability Project Final Report. This report provided the conclusions and results of the licensee's investigation into the location of the missing fuel rod segments and incore detectors, as well as overall control and accountability of SNM at Humboldt Bay. On June 20-24, 2005, a team of inspectors from NRC Headquarters and Region IV conducted the final onsite portion of the special inspection to review the licensee's efforts and details of the Project Final Report. The inspectors concluded that the SNM Control and Accountability Project was generally complete and thorough in its search for the missing fuel segments and incore detectors. The licensee concluded that the most likely location of the missing fuel segments was in the SFP, in an altered configuration while the NRC concluded that the most likely location of the missing fuel segments was at LLW disposal site. Both the licensee and the NRC concluded that the most likely location of the missing incore detectors was a LLW disposal site.

## 3.0 Major Technical or Regulatory Issues

The licensee submitted an ISFSI application in December 2003. The ISFSI dry storage cask will be unique due to the short length of the Humboldt fuel assemblies. Furthermore, the casks will be stored below-grade to accommodate regional seismicity issues, security concerns, and site boundary dose limits. The license for the ISFSI was issued on November 18, 2005. The NRC issued a proposed fine of $96,000 on December 21, 2005, related to the licensee's SNM Control and Accountability Project investigation. Currently, the licensee does not anticipate submitting an LTP until 2009 at the earliest.

## 4.0 Estimated Date For Closure

2011

# Indian Point - Unit 1

## 1.0 Site Identification

Location:          Buchanan, NY
License No.:       DPR-5
Docket No.:        50-3
License Status:    SAFSTOR
Project Manager:   Michael Webb

## 2.0 Site Status Summary

Indian Point-1 operated commercially from August 1962 until October 31, 1974. The plant was shutdown in October 1974 because the emergency core cooling system did not meet regulatory requirements. By January 1976, all spent fuel was removed from the reactor vessel. Some decommissioning work associated with spent fuel storage was performed from 1974 through 1978. The order approving SAFSTOR was issued in January 1996. The PSDAR public meeting was held on January 20, 1999. The licensee plans to decommission Unit 1 with Unit 2, which is currently in operation. The licensee does not plan to begin active decontamination and decommissioning until 2013, when the Indian Point 2 license expires. Since purchasing the Indian Point facility, Entergy has been reviewing its long-term spent fuel storage options for Unit 1, and although it has not finalized its plans it intends to store the Unit 1 fuel in dry storage at the Indian Point Energy Center ISFSI.

## 3.0 Major Technical or Regulatory Issues

None.

## 4.0 Estimated Date For Closure

TBD

# La Crosse

## 1.0 Site Identification

Location: Genoa, WI
License No.: DPR-45
Docket No.: 50-409
License Status: SAFSTOR
Project Manager: Kristina Banovac

## 2.0 Site Status Summary

The La Crosse Boiling Water Reactor (LACBWR) is owned and was operated by the Dairyland Power Cooperative (DPC). LACBWR was a nuclear power plant of nominal 50 MW electrical output which utilized a forced-circulation, direct-cycle boiling water reactor as its heat source. The plant is located on the east bank of the Mississippi River in Vernon County, Wisconsin. The plant was one of a series of demonstration plants funded, in part, by the U.S Atomic Energy Commission (AEC). The nuclear steam supply system and its auxiliaries were funded by the AEC, and the balance of the plant was funded by DPC. The Allis-Chalmers Company was the original licensee; the AEC later sold the plant to DPC and provided them with a provisional operating license.

LACBWR was shut down on April 30, 1987. The SAFSTOR DP was approved August 7, 1991. The DP is considered the PSDAR. The PSDAR public meeting was held on May 13, 1998. DPC has been conducting gradual dismantlement and decommissioning activities. DPC is planning to remove and dispose of its reactor pressure vessel (RPV) in Spring 2007, to take advantage of a disposal window at the Barnwell disposal site. DPC has completed engineering and technical evaluations related to removal and disposal of the RPV. The current decommissioning cost estimate is $84.6 million. The licensee has accumulated approximately $77.3 million in decommissioning funds, as of December 31, 2005.

## 3.0 Major Technical or Regulatory Issues

None.

## 4.0 Estimated Date For Closure

TBD

# Millstone - Unit 1

## 1.0 Site Identification

Location:          Waterford, CT
License No.:       DPR-21
Docket No.:        50-245
License Status:    SAFSTOR
Project Manager:   Alan Wang

## 2.0 Site Status Summary

Millstone - Unit 1 was shut down on November 4, 1995, and transfer of the spent fuel to the pool was completed on November 19, 1995. On July 17, 1998, the licensee decided to cease operations. Certifications per 10 CFR Part 50.82(a) were submitted July 21, 1998. The owner's current plan is to leave the plant in SAFSTOR until the Unit 2 license expires. The owner submitted its required PSDAR on June 14, 1999, and has chosen a combination of the DECON and SAFSTOR options. NRC conducted public meetings in Waterford, CT, on the decommissioning process on February 9, 1999, and on the PSDAR on August 25, 1999. Owner responsibility for the Millstone site was transferred from Northeast Utilities to Dominion Nuclear Connecticut on March 31, 2001. Unit 1 has established a spent fuel pool island including those systems required to support safe storage of spent fuel. The balance of systems not required to support the facility have been abandoned. Irradiated reactor vessel components not able to eventually being disposed of with the reactor vessel have been removed. The reactor cavity and vessel will be drained and abandoned with a radiation shield installed to limit dose to workers.

## 3.0 Major Technical or Regulatory Issues

None.

## 4.0 Estimated Date For Closure

TBD

# Nuclear Ship Savannah

## 1.0 Site Identification

Location:         Washington, DC
License No.:      NS-1
Docket No.:       50-238
License Status:   SAFSTOR
Project Manager:  Al Adams

## 2.0 Site Status Summary

The Nuclear Ship (NS) Savannah was removed from service 1970. The reactor is currently in SAFSTOR. All fuel was removed from the ship in October 1971. The ship is moored in the Maritime Administration Reserve Fleet in the James River, Virginia. As needed, the NS Savannah is towed into dry dock for hull maintenance. Because the reactor is portable, the location of decommissioning has not been determined.

## 3.0 Major Technical or Regulatory Issues

The licensee is exploring the possibility of obtaining funding for total decommissioning and disposal of the NS Savannah. Licensee is planning decommissioning activities. Ship will be drydocked for maintenance in the near future.

## 4.0 Estimated Date For Closure

TBD

# Peach Bottom - Unit 1

## 1.0 Site Identification

Location:            Delta, PA
License No.:         DPR-12
Docket No.:          50-171
License Status:      SAFSTOR
Project Manager:     Kristina Banovac

## 2.0 Site Status Summary

Peach Bottom Atomic Power Station, Unit 1 was a 200 Mwt, high temperature, gas cooled reactor that was operated from June of 1967 to its final shutdown on October 31, 1974. All spent fuel has been removed from the site, and the spent fuel pool is drained and decontaminated. The reactor vessel, primary system piping, and steam generators remain in place.

The facility is currently in a SAFSTOR condition. The PSDAR meeting was held on June 29, 1998. Final decommissioning is not expected until 2034 when Units 2 and 3 are scheduled to shut down. The current decommissioning cost estimate is $146.5 million. The utility will collect approximately $2.6 million annually through 2034 to accumulate sufficient funding. The current amount of decommissioning funds accumulated through December 31, 2005, is $29.8 million.

## 3.0 Major Technical or Regulatory Issues

None.

## 4.0 Estimated Date For Closure

TBD

# Rancho Seco

## 1.0 Site Identification

Location:        Herald, CA
License No.:     DPR-54
Docket No.:      50-312
License Status:  DECON
Project Manager: John Hickman

## 2.0 Site Status Summary

The plant was shutdown in June 1989. The SAFSTOR DP was approved in March 1995. The licensee revised its DP in use an incremental dismantlement approach. Currently, the licensee is dismantling the secondary side of the plant. Wastes generated during decommissioning will be shipped to Envirocare. In July 1999, the owner decided to continue in DECON, with the goal of completing decommissioning by 2008. On October 4, 1991, the owner submitted a site-specific Part 72 ISFSI application using the VECTRA NUHOMS-MP187 dual purpose cask design. The license was granted on June 30, 2000. The owner has transferred all of the spent fuel from the pool to the onsite ISFSI.

The current estimated cost to decommissioning Rancho Seco is $518.6 million (2002 dollars). The licensee has completed dismantlement of the secondary side equipment in the turbine building. Wastes generated during decommissioning are being shipped to Envirocare. The licensee is now dismantling equipment in the auxiliary building. FSSs will then be conducted to verify that structures and open land areas meet the release criteria. Finally, an independent NRC contractor will conduct a verification survey, thereby allowing unrestricted release of the site. After FSSs and NRC verification, individual surveyed structures and open land areas will be released as non-radiologically controlled material for conventional demolition and disposal. Sacramento Municipal Utility District (SMUD) will maintain control over the site until termination of its 10CFRPart 50 license.

## 3.0 Major Technical or Regulatory Issues

SMUD submitted a LTP for NRC review on April 12, 2006. The LTP is currently under review.

## 4.0 Estimated Date For Closure

2008

# San Onofre - Unit 1

## 1.0 Site Identification

Location: San Clemente, CA
License No.: DPR-13
Docket No.: 50-206
License Status: DECON
Project Manager: Jim Shepherd

## 2.0 Site Status Summary

The San Onofre Nuclear Generating Station (SONGS), operated by Southern California Edison (SCE) is approximately 100 km (60 mi) south of Los Angeles, 6.5 km (4 mi) south of San Clemente, CA. It is located between I-5 and the Pacific Ocean, within the boundary of the Camp Pendelton military reserve. SONGS-1, a Westinghouse 3-loop pressurized water reactor constructed by Bechtel and rated at 1347 MWt, began commercial operation on January 1, 1968. It ceased operation on November 30, 1992. Defuelling was completed on March 6, 1993. The licensee has transferred Unit 1 spent fuel to an onsite generally licensed ISFSI.

Significant dismantlement of Unit 1 is currently underway. Units 2 and 3 are expected to operate until approximately 2022. The licensee has completed demolition of the Unit 1 Emergency Diesel Generator building, Control Building, and Administration Building. Dismantlement and removal of the electrical generator and main turbine is also complete. The licensee has completed RPV internal segmentation and cutup. The reactor internals abrasive cutting media has been sent offsite for disposal. Most of the Containment Sphere Enclosure Building has been dismantled and most of the large reactor system components have been removed including the RPV, pressurizer and steam generators. The remaining structure inside containment is being removed, and the turbine building is being removed down to about 10 feet below grade. The steam generators and pressurizer have been shipped to disposal. The licensee was unable to make arrangements for shipping the RPV to disposal because of the size and weight of the vessel and shipping package. The licensee plans to store the vessel onsite for the foreseeable future, as long as licensed activities are ongoing.

On November 3, 1994, SCE submitted a proposed DP to place SONGS-1 in SAFSTOR until the shutdown of Units 2 and 3, at the end of their licenses in 2022. On December 15, 1998, following a change in NRC decommissioning regulations, SCE submitted a PSDAR for SONGS-1 to commence DECON in 2000. Since that time, it has been actively decommissioning the facility.

## 3.0 Major Technical or Regulatory Issues

SCE plans to leave the off-shore portions of the Unit 1 intake and outlet pipes in place, under the Pacific Ocean seabed, and release them for unrestricted use and terminate the lease it has from California. This would constitute a partial site release prior to submission of the LTP in accordance with 10 CFR 50.83. SCE has submitted an Environmental Report to the State. SCE discussed its plans with NRC during 2006 and plans to submit a request to NRC for release of this system in 2007. During 2006, SCE will survey the intake "boxes" for this piping, conduct remediation if deemed necessary, and fill them with grout.

SCE has elected to leave the below-grade portions of the turbine building in place after grouting expansion joints and embedded pipes. Because SCE has not submitted an LTP for this unit, it is not known if the surveys done on these areas prior to grouting will meet NRC requirements for FSSs at the time of request for license termination. The PSDAR states all equipment and structures from Unit 1 will be removed from the site at the time of license termination, but SCE has stated it may reconsider this later, and possibly leave some of the below-grade structures in place. Current survey data may not support this option, in which case additional surveys, e.g. of the embedded piping, may be necessary to implement it.

## 4.0 Estimated Date For Closure

2045

# Three Mile Island - Unit 2

## 1.0 Site Identification

Location:           Middletown, PA
License No.:        DPR-73
Docket No.:        50-320
License Status:    SAFSTOR
Project Manager:  Kristina Banovac

## 2.0 Site Status Summary

The Three Mile Island, Unit 2 (TMI-2) operating license was issued on February 8, 1978, and commercial operation was declared on December 30, 1978. On March 28, 1979, the unit experienced an accident which resulted in severe damage to the reactor core. TMI-2 has been in a non-operating status since that time. The licensee conducted a substantial program to defuel the reactor vessel and decontaminate the facility. All spent fuel has been removed except for some debris in the nuclear steam supply system. Plant defueling was completed in April 1990. The removed fuel is currently in storage at Idaho National Engineering and Environmental Laboratory, and DOE has taken title and possession of the fuel. TMI-2 has been decontaminated to the extent the plant is in a safe, inherently stable condition suitable for long-term management. This long-term management condition is termed post-defueling monitored storage, which was approved in 1993. There is no significant dismantlement underway. The plant shares equipment with the operating TMI - Unit 1. TMI-1 was sold to Amergen in 1999. GPU Nuclear retains the license for TMI-2 and is owned by First Energy Nuclear Operating Corporation. GPU Nuclear contracts with Amergen for maintenance and surveillance activities. The licensee plans to actively decommission TMI-2 in parallel with the decommissioning of TMI-1.

The current radiological decommissioning cost estimate is $753 million and $25 million for non-radiological funds, as of December 31, 2005. The current amount in the decommissioning trust fund is $494 million, as of December 31, 2005.

## 3.0 Major Technical or Regulatory Issues

None.

## 4.0 Estimated Date For Closure

2014

# Vallecitos Boiling Water Reactor

## 1.0 Site Identification

Location:          Sunol, CA
License No.:       DPR-1
Docket No.:        50-18
License Status:    SAFSTOR
Project Manager:   Marvin Mendonca

## 2.0 Site Status Summary

The VBWR was shutdown in 1963 and NRC issued a possession only license in 1965. The license was renewed in 1973 and the license has remained effective under the provisions of 10 CFR 50.51(b). The facility has been maintained in SAFSTOR condition. The site has an operating research reactor, and has hot cells that are used for power reactor fuel post irradiation examination. The licensee plans to maintain the facility in SAFSTOR until ongoing nuclear activities are terminated and the entire site can be decommissioned. General Electric has a self-guarantee instrument. The spent fuel has been removed from the site.

## 3.0 Major Technical or Regulatory Issues

None.

## 4.0 Estimated Date For Closure

TBD

# Yankee Rowe

## 1.0 Site Identification

| | |
|---|---|
| Location: | Rowe, MA |
| License No.: | DPR-3 |
| Docket No.: | 50-29 |
| License Status: | DECON |
| Project Manager: | John Hickman |

## 2.0 Site Status Summary

The plant was permanently shut down on October 1, 1991. The DECON DP was approved in February 1995, and the plant is undergoing dismantlement. The steam generators were shipped to the Barnwell, North Carolina LLW facility in November 1993. The reactor vessel was shipped to Barnwell in April 1997. The owner has removed all of the primary systems, secondary side components, and switch yard equipment from the site. The plant is about 80 percent dismantled. The containment and other major structures remain. The owner has completed construction of an onsite ISFSI. An LTP was submitted in May 1997, and a public meeting was held to discuss the LTP in January 1998. A public hearing was requested on the LTP but was cancelled after the owner withdrew the plan in May 1999. The licensee resubmitted portions of the revised LTP in November 2003, and the remainder in March 2004. The staff completed its review in April 2005. All of the fuel from the spent fuel pool has been transferred to the onsite ISFSI.

## 3.0 Major Technical or Regulatory Issues

FSS reports and the Groundwater Compliance Plan for License Termination are currently under review.

## 4.0 Estimated Date For Closure

2007

# Zion - Units 1 & 2

## 1.0 Site Identification

Location:          Warrenville, IL
License No.:       DPR-39 & DPR-48
Docket No.:        50-295 & 50-304
License Status:    SAFSTOR
Project Manager:   John Hickman

## 2.0 Site Status Summary

Zion Units 1 and 2 were permanently shut down on February 13, 1998. The fuel was transferred to the spent fuel pool, and the owner submitted the certification of fuel transfer on March 9, 1998. The owner has converted the turbine-generators into synchronous condensers and has isolated the spent fuel pool within a fuel building "nuclear island." The plant has been placed in SAFSTOR, where it will remain until about 2013 when the decommissioning trust fund will be sufficient to conduct DECON activities. The owner submitted the PSDAR, site-specific cost estimate, and fuel management plan on February 14, 2000.

Zion's operating license was issued April 6, 1973, for Unit I, and November 14, 1973, for Unit 2. Commercial operations achieved, December 1973, for Unit 1 and September 1974, for Unit 2. Reactor shutdown occurred on February 21, 1997, for Unit 1, and September 19, 1996, for Unit 2. All fuel was removed from the reactor and placed in the spent fuel pool on April 27, 1997, for Unit 1, and February 25, 1998, for Unit 2. On January 14, 1998, the Unicom Corporation and ComEd Boards of Directors authorized the permanent cessation of operations at ZNPS for economic reasons. The NRC approved the transfer of the facility licenses from ComEd Company to Exelon on January 12, 2001.

The SAFSTOR approach is the intended decommissioning method to be utilized for Zion which involves removal of all radioactive material from the site following a period of dormancy. Preparations for decontamination and dismantlement are scheduled to commence at the original license expiration date for ZNPS Unit 2 on November 14, 2013. FSS and license termination is currently planned for 2025 - 2026.

## 3.0 Major Technical or Regulatory Issues

None.

## 4.0 Estimated Date For Closure

2026

# Appendix B

# Site Summaries for
# Research and Test Reactors

# Cornell University - TRIGA

## 1.0 Site Identification

Location:          Ithaca, NY
License No.:       R-80
Docket No.:        50-157
License Status:    DECON
Project Manager:   Daniel E. Hughes

## 2.0 Site Status Summary

Cornell University submitted a request for approval of a decommissioning amendment on August 22, 2003, for R-80 which is a 500 kR TRIGA reactor. The decommissioning of the R-80 reactor will be concurrent with the decommissioning of Cornell's zero power reactor (R-89). There is no fuel on site for this reactor.

## 3.0 Major Technical or Regulatory Issues

None.

## 4.0 Estimated Date For Closure

2007

# Cornell University - ZPR

## 1.0 Site Identification

Location:          Ithaca, NY
License No.:     R-89
Docket No.:     50-97
License Status:  DECON
Project Manager:  Daniel E. Hughes

## 2.0 Site Status Summary

Cornell University submitted a request for approval of a decommissioning amendment on August 22, 2003, for R-80 which is a 500 kilowatt (kW) TRIGA reactor. The decommissioning of the R-80 reactor will be concurrent with the R-89 reactor. There is no fuel on site for this reactor.

## 3.0 Major Technical or Regulatory Issues

None.

## 4.0 Estimated Date For Closure

2007

# Ford Nuclear Reactor

## 1.0 Site Identification

Location:          Ann Arbor, MI
License No.:       R-28
Docket No.:        50-2
License Status:    DECON
Project Manager:   Patrick J. Isaac

## 2.0 Site Status Summary

The construction of the Ford Nuclear Reactor (FNR), located in the Phoenix Memorial Laboratory (PML) began in 1956. The PML is a four story, reinforced concrete building. The FNR and PML are located on the North Campus of the UM in Ann Arbor, Michigan.

In 1957 the FNR went critical. The reactor is a 2 MW, open pool reactor facility. The decommissioning plan (DP) for the University of Michigan FNR was approved on June 26, 2006. It is de-fueled.

## 3.0 Major Technical or Regulatory Issues

None.

## 4.0 Estimated Date For Closure

TBD

# General Atomics - TRIGA Mark F

## 1.0 Site Identification

Location:        San Diego, CA
License No.:     R-67
Docket No.:      50-163
License Status:  DECON
Project Manager: Al Adams

## 2.0 Site Status Summary

Decommissioning activities at General Atomics (GA) are currently on hold pending the return of fuel to the U.S. Department of Energy (DOE). The licensee has dismantled the Mark F reactor to the extent possible given the storage of fuel.

## 3.0 Major Technical or Regulatory Issues

DOE has refused to take the reactor fuel. DOE is concerned that accepting fuel from GA could impact legal issues surrounding DOE acceptance of fuel from the nuclear power industry.

## 4.0 Estimated Date For Closure

TBD

# General Atomics - TRIGA Mark I

## 1.0 Site Identification

Location:          San Diego, CA
License No.:       R-38
Docket No.:        50-89
License Status:    DECON
Project Manager:   Al Adams

## 2.0 Site Status Summary

Decommissioning activities at GA are currently on hold pending return of fuel to DOE. The licensee has dismantled the Mark I reactor to the extent possible given the storage of fuel. To complete decommissioning activities on the Mark I reactor, the licensee needs to dismantle parts of the building in which the Mark I and Mark F reactors are located. These activities are on hold until fuel is returned to DOE.

## 3.0 Major Technical or Regulatory Issues

DOE has refused to take the reactor fuel. DOE is concerned that accepting fuel from GA could impact legal issues surrounding DOE acceptance of fuel from the nuclear power industry.

## 4.0 Estimated Date For Closure

TBD

# General Electric Co. - GETR

## 1.0 Site Identification

Location:           Sunol, CA
License No.:        TR-1
Docket No.:         50-70
License Status:     SAFSTOR
Project Manager:    Marvin Mendonca

## 2.0 Site Status Summary

NRC issued a possession-only license for GETR on February 5, 1986. The license was renewed on September 30, 1992, to expire in 2016. The facility has been maintained in SAFSTOR condition. The site has an operating research reactor, and has hot cells that are used for power reactor fuel post irradiation examination. The licensee plans to maintain the facility in SAFSTOR until ongoing nuclear activities are terminated and the entire site can be decommissioned.

## 3.0 Major Technical or Regulatory Issues

None.

## 4.0 Estimated Date For Closure

TBD

# General Electric Co. - VESR

## 1.0 Site Identification

Location:          Sunol, CA
License No.:       DR-10
Docket No.:        50-183
License Status:    SAFSTOR
Project Manager:   Marvin Mendonca

## 2.0 Site Status Summary

On April 15, 1970, the U.S. Nuclear Regulatory Commission (NRC) authorized the licensee to possess but not operate the reactor. The license was renewed on June 11, 1976, to expire in 2016. The facility has been maintained in SAFSTOR condition. The facility is next to the Vallecitos Boiling Water Reactor which is also in SAFSTOR. The licensee plans to maintain the facility in SAFSTOR until other ongoing nuclear and radioactive activities are also to be decommissioned to provide an integrated site decommission.

## 3.0 Major Technical or Regulatory Issues

None.

## 4.0 Estimated Date For Closure

TBD

# NASA - MOCKUP

## 1.0 Site Identification

| | |
|---|---|
| Location: | Cleveland, OH |
| License No.: | R-93 |
| Docket No.: | 50-185 |
| License Status: | DECON |
| Project Manager: | Patrick J. Isaac |

## 2.0 Site Status Summary

The NASA Plum Brook Reactor Facility (PBRF) is located within a fenced area in the northern portion of NASA's Plum Brook Station. The Plum Brook Station is located about 6-km (4-mi) south of Sandusky, Ohio. NASA currently has 10 CFR Part 50 facility licenses to "possess but not operate" two reactors within the PBRF. NRC license TR-3 is for the 60-megawatt research test reactor, constructed for testing materials to be used in space program applications. NRC license R-93 is for the 100-kilowatt swimming-pool type Mock-Up Reactor (MUR). Upon approval the DP, these two licenses were amended on March 20, 2002 to allow decommissioning of the facility. The PBRF operated from 1961 to 1973. The facility is to be decommissioned, with the end objective being removal and disposal of remaining radioactive materials, release of the 11-ha (27-acre) facility for unrestricted use, and termination of the NRC licenses. The radiological criteria for license termination to allow unrestricted use are set forth in 10 CFR Part 20, Subpart E. After many years of little to no activities at the Plum Brook reactor site, decommissioning is well underway. NRC approved the DP in March 2002, and in November 2002, NASA conducted the first reactor tank entry in 30 years. In August 2003, NASA began taking important steps in removing the reactor internals and segmenting the reactor tank for shipment to Barnwell, SC. NASA plans to complete decommissioning by 2010.

## 3.0 Major Technical or Regulatory Issues

NASA recently discovered radioactive contamination on and off the NASA Plum Brook test reactor site near Sandusky, Ohio. The material was identified as cesium-137 and cobalt-60 in a drainage ditch leaving their property, and in Plum Brook approximately one mile downstream towards Lake Erie. The radioactive materials are likely the result of reactor operations which ended in about 1973. Sediment samples identified up to 38 pCi/l cesium-137 (background is about 1 pCi/g).

## 4.0 Estimated Date For Closure

2010

# NASA - Plum Brook

## 1.0 Site Identification

Location:           Cleveland, OH
License No.:        TR-3
Docket No.:         50-30
License Status:     DECON
Project Manager:    Patrick J. Isaac

## 2.0 Site Status Summary

The NASA PBRF is located within a fenced area in the northern portion of NASA's Plum Brook Station. The Plum Brook Station is located about 6-km (4-mi) south of Sandusky, Ohio. NASA currently has 10 CFR Part 50 facility licenses to "possess but not operate" two reactors within the PBRF. NRC license TR-3 is for the 60-megawatt research test reactor, constructed for testing materials to be used in space program applications. NRC license R-93 is for the 100-kilowatt swimming-pool type Mock-Up Reactor (MUR). Upon approval the DP, these two licenses were amended on March 20, 2002 to allow decommissioning of the facility. The PBRF operated from 1961 to 1973. The facility is to be decommissioned, with the end objective being removal and disposal of remaining radioactive materials, release of the 11-ha (27-acre) facility for unrestricted use, and termination of the NRC licenses. The radiological criteria for license termination to allow unrestricted use are set forth in 10 CFR Part 20, Subpart E. After many years of little to no activities at the Plum Brook reactor site, decommissioning is well underway. NRC approved the DP in March 2002, and in November 2002, NASA conducted the first reactor tank entry in 30 years. In August 2003, NASA began taking important steps in removing the reactor internals and segmenting the reactor tank for shipment to Barnwell, SC. NASA plans to complete decommissioning by 2010.

## 3.0 Major Technical or Regulatory Issues

NASA recently discovered radioactive contamination on and off the NASA Plum Brook test reactor site near Sandusky, Ohio. The material was identified as cesium-137 and cobalt-60 in a drainage ditch leaving their property, and in Plum Brook approximately one mile downstream towards Lake Erie. The radioactive materials are likely the result of reactor operations which ended in about 1973. Sediment samples identified up to 38 pCi/l cesium-137 (background is about 1 pCi/g).

## 4.0 Estimated Date For Closure

2010

# University of Buffalo

## 1.0 Site Identification

Location:              Buffalo, NY
License No.:           R-77
Docket No.:            50-57
License Status:        Possession Only License
Project Manager:   Daniel E. Hughes

## 2.0 Site Status Summary

License R-77 was amended June 6, 1997, for possession only. There is no fuel on site. A DP has not been submitted.

## 3.0 Major Technical or Regulatory Issues

None.

## 4.0 Estimated Date For Closure

TBD

# University of Illinois

## 1.0 Site Identification

Location:          Urbana, IL
License No.:       R-111
Docket No.:        50-151
License Status:    DECON
Project Manager:   Al Adams

## 2.0 Site Status Summary

Reactor fuel has been removed from the facility. The licensee has an approved DP that allows limited decommissioning activity.

## 3.0 Major Technical or Regulatory Issues

None.

## 4.0 Estimated Date For Closure

TBD

# University of Washington

## 1.0 Site Identification

Location:          Seattle, WA
License No.:       R-73
Docket No.:        50-139
License Status:    DECON
Project Manager:   Al Adams

## 2.0 Site Status Summary

Decommissioning activities started in March 2006 and are well underway. An Order was issued by the staff to allow 10CFR50.59-type changes to be made to the DP.

## 3.0 Major Technical or Regulatory Issues

None.

## 4.0 Estimated Date For Closure

2007

# Veterans Administration

## 1.0 Site Identification

Location:           Omaha, NE
License No.:        R-57
Docket No.:         50-131
License Status:     Possession Only
Project Manager:    Al Adams

## 2.0 Site Status Summary

TRIGA Mark-1 research reactor was operated at 20 kW thermal power from 1959 until November 5, 2001. All fuel has been removed from the site. A DP was submitted 9/21/04. The DP is currently under staff review.

## 3.0 Major Technical or Regulatory Issues

None.

## 4.0 Estimated Date For Closure

TBD

# Westinghouse

## 1.0 Site Identification

Location:         New Stanton, PA
License No.:      TR-2
Docket No.:       50-22
License Status:   DECON
Project Manager:  Patrick J. Isaac

## 2.0 Site Status Summary

The TR-2 License was amended in March 1963 to allow possession, but not use of the reactor. The Westinghouse test reactor, located on the Waltz Mill site, is undergoing decommissioning in accordance with the DP which was approved in September 1998. CBS (formerly Westinghouse Electric Corporation), which operated the Waltz Mill Facility, was the licensee of the TR-2 and SNM-770. Radiological contamination in soil and groundwater exist on a portion of the site as a result of the clean-up activities following a 1961 incident at the test reactor, waste segregation activities, and nuclear laundry services. Significant contamination is also present in retired facilities (hot cells, hot cell support rooms, and a section of the fuel transfer canal) within one of the site buildings. Contaminants are primarily strontium-90 (Sr-90) and cesium-137 (Cs-137), with lesser quantities of mixed fission, activation products, and trace levels of transuranic radionuclides. The TR-2 DP required removal of designated portions of the shutdown reactor as necessary and sufficient to terminate the Part 50 portion of the license. At that point, the remaining residual radioactive materials would be transferred to SNM-770 where they would continue to be controlled under that license. In March 1999, Viacom acquired the TR-2 license and a new company, Westinghouse Electric Company, LLC, (Westinghouse) became the holder of the SNM-770 License. Westinghouse and Viacom entered into a project management agreement whereby Westinghouse agreed to act as Viacom's decommissioning project manager for the TR-2 reactor. The pressure vessel and pressure vessel internals have been removed in accordance with the DP, as well as the biological shield that needed to be removed in order to remove the pressure vessel. Two provisions of the DP still need to be accomplished: determining the residual radioactivity remaining in situ and preparing the necessary amendments for and requesting the transfer to the SNM-770 license.

## 3.0 Major Technical or Regulatory Issues

Westinghouse is refusing to accept the transfer to the SNM-770 license. Viacom filed a 10 CFR 2.206 petition in which it alleges that Westinghouse is in violation of 10 CFR 50.5, Deliberate Misconduct. Westinghouse claims that Viacom did not perform all the actions required prior to the transfer. NRC issued a Director's Decision concluding that Westinghouse was not in violation of 10 CFR 50.5 on August 26, 2003. Viacom and Westinghouse are currently engaged in a commercial dispute and are under arbitration to resolve the disputed issues.

## 4.0 Estimated Date For Closure

TBD

# Appendix C

# Site Summaries for Current Complex Decommissioning Sites

# AAR Manufacturing, Inc.

## 1.0 Site Identification

Location:           Livonia, MI
License No.:        STB-362
Docket No.:         40-0235
License Status:     Terminated
Project Manager:    Kristina Banovac

## 2.0 Site Status Summary

The AAR Manufacturing, Inc. (AAR) site, located in Livonia, Michigan, was formerly owned by Brooks & Perkins, Inc. (B&P), a licensee of the U.S. Atomic Energy Commission (AEC). AEC Source Material License D-547 was issued to B&P on January 17, 1957, and then was superceded by License No. STB-0362 on August 10, 1961. AEC terminated license STB-0362 on May 17, 1971. In 1981, AAR purchased B&P and obtained the property. Thorium contaminated surface and subsurface soil has been identified at several locations on the site. The site was added to the Site Decommissioning Management Plan (SDMP) list in August 1994.

AAR submitted its final remediation plan (RP) on October 14, 1997, and NRC approved the RP on May 22, 1998. In November 1998, AAR completed additional site characterization and identified large volumes of soil that contained thorium in concentrations exceeding the approved cleanup criterion. In September 1999, AAR effectively withdrew its approved RP and proposed using "unimportant quantities of source material" (0.05 percent by weight source material), as defined in 10 CFR 40.13(a), as a decommissioning criterion. After staff consultation with the Commission on this policy issue, NRC informed AAR that the revised remediation approach was not acceptable, by letter dated August 9, 2002. AAR is currently proposing unrestricted use of the eastern portion of the site and restricted use of the western portion of the site. AAR plans to enter into a settlement agreement with the NRC on the restrictions and controls needed for restricted use. The agreement would include using a restrictive covenant that would outline the restrictions on the use of the site, such as prohibiting farming or developing residential properties on the site. The agreement would allow NRC or the local and State government to enforce the controls. The cost of decommissioning is unknown at this time.

## 3.0 Major Technical or Regulatory Issues

AAR is not a licensee. AAR believes it should not be responsible for the cost of site remediation, since it was not directly responsible for the contamination onsite. The staff is currently working with AAR to resolve technical issues associated with its dose assessment. AAR consultants are currently considering the staff's issues with the dose assessment and are determining the impacts on the estimate of soil remediation needed. Elevated levels of thorium have also been identified along the fence separating AAR and CSX Transportation, Inc. (CSX). Although contamination appears to be very limited, there is the potential that financial responsibility for the contamination on CSX property may become an issue. No remediation has been performed by CSX.

## 4.0 Estimated Date For Closure

09/01/08

# ABB Prospects, Inc. (Formerly C.E. Windsor)

## 1.0 Site Identification

Location:             Windsor, CT
License No.:          06-00217-06; SNM-1067
Docket No.:           030-03754; 070-01100
License Status:       Possession Only
Project Manager:      Laurie Kauffman

## 2.0 Site Status Summary

The ABB Prospects, Inc., (formerly Combustion Engineering-Windsor) site consists of soils, and building and equipment surfaces contaminated with uranium and by-product material from operations that occurred from the late 1950s until 2001. A site-wide decommissioning plan (DP) was received by NRC on April 7, 2003, and a revised DP, which includes dose modeling information, was received on October 15, 2003. On June 1, 2004, the license was amended to incorporate the DP. On June 30, 2004, the licensee submitted the site-wide final status survey (FSS) plan. Under the current License 06-00217-06, the licensee removed interior systems, components, ducts, piping, conduits from Building Complexes 2, 5, and 17. Equipment and material are being released from the site using Regulatory Guide 1.86 criteria as permitted by the current license. The present license also permits the licensee to demolish the buildings of Building Complexes 2, 5, and 17 down to grade level only. FSS and sample collection of Building Complexes 2, 5, and 17 began in 2002 and decontamination and demolition of the buildings was completed in July 2005. The waste materials were shipped off site to a licensed disposal facility. Dismantlement of the waste water treatment system and the health physics offices and laboratory are in progress. The FSS reports for the Building 2, 5/6A, and 17 Complexes were submitted to the NRC on October 31, 2005, February 7, 2006, and May 2, 2006, respectively. The NRC completed its acceptance review of each FSS report and determined that the submissions were sufficiently complete for NRC to initiate a detailed technical review. The technical review is scheduled for completion by December 31, 2006. On June 30, 2005, the licensee submitted an application to renew the NRC SNM-1067 License, per the requirements for timely renewal. Also, the licensee wishes to maintain the SNM-1067 license if any residual uranium is detected on surfaces and/or in soils being remediated by the United States Army Corps of Engineers (USACE), who is responsible for remediating certain areas of the site under the Formerly Utilized Sites Remedial Action Program (FUSRAP). NRC staff completed the renewal on January 18, 2006. Once USACE has remediated the FUSRAP areas, ABB Prospects, Inc. must demonstrate to the NRC that the entire site meets the criteria for unrestricted release in accordance with Subpart E of 10 CFR Part 20. The licensee estimates the cost of decommissioning to be approximately $2.6 million, based on the licensee's decommissioning funding plan dated December 2003, for License 06-00217-06.

ABB Prospects, Inc. manufactured nuclear fuels, and at various times, was used to conduct and support research and development. The site's activities started in 1955 with an AEC contract to begin research, development, and manufacturing of nuclear fuel for the United States Navy. These activities included the construction, testing, and operation of a U.S. Naval test reactor.

## 3.0 Major Technical or Regulatory Issues

There are no technical or regulatory issues. The site participates in the U.S. Environmental Protection Agency (EPA) Voluntary corrective Action Program and last met with EPA on March 11, 2005.

## 4.0 Estimated Date For Closure

12/01/2007

# Babcock & Wilcox (Shallow Land Disposal Area)

## 1.0 Site Identification

Location: Vandergrift, PA
License No.: SNM-2001
Docket No.: 07003085
License Status: Possession Only
Project Manager: Amir Kouhestani

## 2.0 Site Status Summary

The Babcock & Wilcox (BWXT) shallow land disposal area (SLDA) site is situated in Parks Township, PA, and consists of 10 trenches that were used to dispose of wastes, scrap, and trash from a nearby nuclear fuel fabrication facility in Apollo, PA. Principal radioactive contaminants at the site are natural uranium, enriched uranium, and DU, and lesser quantities of Am-241, plutonium, and thorium. In 1970, Nuclear Materials and Equipment Company (NUMEC), a former site licensee, ceased the use of SLDA for radioactive waste disposal. In 2000, the site was designated as a FUSRAP site. USACE is responsible for administrating site cleanup. The SLDA site will be decommissioned by USACE consistent with the USACE-NRC memorandum of understanding (MOU), and the NRC radiological criteria for unrestricted use. USACE will issue a Record of Decision (ROD) in place of a DP. In September 2005, NRC responded to the licensee's request to continue with its current license commitments beyond its license expiration date of October 31, 2005.

The SLDA was created for the disposal of uranium-contaminated waste generated by NUMEC, between 1961 and 1970. NUMEC's mission was to convert enriched uranium to naval reactor fuel. NUMEC operated the nearby Apollo nuclear fuel fabrication facility in the late 1950s. The waste from this facility was disposed of in trenches at the SLDA in accordance with the AEC regulation in effect at the time, 10 CFR 20.304.

## 3.0 Major Technical or Regulatory Issues

In the event that USACE does not complete the Congressionally mandated site remediation, NRC staff anticipates that BWXT-SLDA may request license termination, with restrictions on future land use. The Commonwealth of Pennsylvania Department of Environmental Protection (PADEP), the cognizant state agency responsible for radiation protection, has stated that it will not assume responsibility for the site (i.e., become the institutional control authority) if the site is decommissioned with land-use restrictions. Due to quantities of special nuclear materials at the SLDA exceeding the NRC Unity Rule, the site will not be transferred to Commonwealth of Pennsylvania after it becomes an Agreement State. In March 2006, in compliance with the implementing requirements of the Comprehensive Environmental Response, Compensation, and Liability Act of 1980 (CERCLA), USACE issued a draft Final Feasibility Study (FS) report. The draft FS report discusses USACE's remedial action objectives for the SLDA site and presents an analysis of USACE's proposed remedial alternatives. In the draft FS report, USACE discusses NRC's restricted use criteria for site decommissioning and license termination as the preferred criteria and alternative. In May 2006, NRC staff commented on the draft FS. Also, in May, PADEP commented on the FS proposed remedial alternative involving construction of a disposal cell on site. PADEP stated that construction of an on-site disposal cell is not allowed under Pennsylvania law. USACE's tentative plan called for a combined submittal of the Final FS and its Proposed Plan (PP) to NRC and PADEP by end of August 2006. NRC has not yet received USACE's final FS/PP report. There is significant public and Congressional interest in the site. No financial assurance issues have been identified at this time.

## 4.0 Estimated Date For Closure

10/01/2009

# Battelle Columbus Laboratories

## 1.0 Site Identification

Location:          Columbus, OH
License No.:       SNM-00007
Docket No.:        070-00008
License Status:    Decon
Project Manager:   George M. McCann

## 2.0 Site Status Summary

Battelle Memorial Institute's (BMI's) base license authorized two sites, the King Avenue site, which was located in Columbus, Ohio, and the West Jefferson Nuclear Sciences site, which was located in West Jefferson, Ohio. In 1977, following decommissioning of the Battelle Research Reactor, which was located at the West Jefferson site, the portion of the license authorizing special nuclear material fuels research at the West Jefferson site was converted from operation to possession only. However, the licensee continued to perform research with byproduct material at its Columbus, Ohio site. In December 1993, the NRC issued a letter approving Battelle's DP, which authorized decommissioning activities at the Columbus West Jefferson sites. Decommissioning oversight for the Columbus site was transferred to the State of Ohio, when the State became a NRC Agreement State in 1999.

The entire West Jefferson site comprises a 1,183-acre tract. The Nuclear Sciences Area occupies an 11-acre fenced area in the northern portion of the West Jefferson site. Outside of the fenced area, several active and abandoned filter beds, and part of the site sanitary sewer systems were also included in the project.

BMI is the licensee for the project. The U.S. Department of Energy (DOE), who is funding the cleanup of the site, decided in 2003 to discontinue using BMI as the prime contractor for performing decontamination activities at the West Jefferson site. Instead, DOE chose to use an independent contractor who reports directly to DOE. Remediation and decontamination activities are conducted under the authority of the BMI license, with BMI continuing to be responsible for site cleanup and regulatory consequences of the contractor's activities. The new contractor resumed decommissioning activities in March 2004.

In February 2006, the contractor completed all building demolition, subsurface foundation excavation and remediation, and back-filling operations at the former North Nuclear Sciences Site. FSSs were completed during March 2006. The Oak Ridge Institute for Science and Education (ORISE), under contract to DOE, also completed site verification surveys. On August 3, 2006, the licensee filed a license amendment for the termination of its license. NRC expects to complete its review of BMI's FSSs by mid-October 2006.

## 3.0 Major Technical or Regulatory Issues

In July 2006, BMI decided to install monitoring wells to sample all potentially impacted groundwater. By September 2006, BMI had installed 22 monitoring wells, with 9 wells on the former licensed site and the remainder placed in areas off-site.

The licensee identified a few groundwater monitoring wells, which exceeded the EPA's maximum concentration limits (MCLs) for strontium 90, and gross and beta levels. Staff has determined that a level 2 notification to the EPA is necessary. It is anticipated that a consultation letter to EPA regarding BMI's groundwater will be completed by mid-November 2006.

## 4.0 Estimated Date For Closure

11/06

# Cabot Performance Materials, Inc.

## 1.0 Site Identification

| | |
|---|---|
| Location: | Boyertown, PA |
| License No.: | SMC-1562 |
| Docket No.: | 040-09027 |
| License Status: | Possession-only |
| Project Manager: | Ted Smith |

## 2.0 Site Status Summary

Contamination at the Cabot Performance Materials, Inc. (Cabot) site consists of surface and subsurface uranium and thorium contamination, in the form of slag. Ground water contamination has not been identified at the site. The March 2000 DP, as supplemented in November 2002, proposes unrestricted release of the site in its current condition. NRC staff issued a request for additional information (RAI) in March 2003, for additional information regarding site characterization, source term modeling, and previously unconsidered aspects of meeting the "as low as is reasonably achievable" requirements of the License Termination Rule (LTR) at the site. The licensee provided a proposed conceptual approach to resolving NRC questions in 2004. The conceptual approach includes emplacement of a riprap cover as an engineered barrier.

Slag at the Reading site was generated from the processing of iron and tin ores for tantalum in 1967 and 1968. The slag was disposed in a pile on a hillside at the edge of the site. Additional source material was placed on the pile when the process building was decontaminated in 1977 and 1978, and contaminated slag from the Canton Yards site in Baltimore, Maryland, was placed on the pile. The pile encompasses approximately 5100 cubic meters (180,000 cubic feet). As described by Cabot, the average contamination levels are 45 pCi/g thorium-232 and progeny, and 30 pCi/g uranium-238 and progeny. The residual radioactivity consists of surface and subsurface uranium and thorium contamination, in the form of slag, in the slag pile and in a road and railroad right of way at the toe of the slope. The first DP submittal was in August 1998. For this first DP submittal, NRC noticed the receipt of the DP and provided an opportunity for a hearing in the Federal Register on October 28, 1998. Two parties [Reading Redevelopment Authority/City of Reading, and Jobert Inc./Metals Trucking Inc. (owner of the site at the time of filing)] petitioned for a hearing. In March 2000, the City of Reading took title to the property. In May 2000, the Jobert Inc./Metals Trucking Inc hearing request was vacated. Several months of private negotiations between the City of Reading and Cabot concluded with the City's request to withdraw their hearing request. The court vacated the City of Reading's hearing request in October 2000. A town meeting was held in January 2003; issues identified for follow on activities were related to non-radiological aspects of the area, and PADEP is addressing. The City of Reading is interested in two Brownfield's redevelopment projects in the immediate vicinity of the site. Currently, Cabot proposes to leave the material in place, using criteria in the LTR, with a rip-rap erosion barrier. The licensee submitted Rev. 4 of their DP in August 2006. The NRC is reviewing Cabot's DP.

## 3.0 Major Technical or Regulatory Issues

The licensee's revised DP proposes use of an engineered barrier to prevent erosion at the site. No major financial assurance issues are associated with this site. A potential financial assurance concern would arise if off-site disposal were required. Public interest in the decommissioning activities at the site was increasing in late 2002, but has since subsided. The City of Reading is conducting a major revitalization project of the city, which includes reutilization of land areas near the site.

## 4.0 Estimated Date For Closure

10/07

# Curtiss-Wright Cheswick

## 1.0 Site Identification

Location: Cheswick, PA
License No.: SNM-1120
Docket No.: 070-01143
License Status: Possession-only
Project Manager: Mark Roberts

## 2.0 Site Status Summary

The Curtiss-Wright Electro-Mechanical Corporation (formerly Westinghouse Government Services) facility is a multi-building complex situated on 110 acres near Cheswick, PA. The Allegheny River is approximately one mile south of the facility. Past commercial and government fuel fabrication operations with low-enriched and highly enriched uranium have left a legacy of contamination in, on, around, and under some of the site buildings. Buildings 4, 5, 5A, 5B, 5C, and 5F show some fixed contamination in generally inaccessible areas or contamination in drain lines. Fuel fabrication activities ceased in the early 1970s; however, the remaining contamination in many locations is deposited in areas that are inherent to the design and structure of the facility (i.e., exterior load-bearing walls, structural steel supports, drain lines beneath significant equipment, and roof supports). The facility has two other NRC radioactive materials licenses for contaminated motor servicing and radiography. Buildings 4 and 5 are presently being used for manufacturing and support functions for its pump and motor operations. Contaminated debris was uncovered in 1984 in a former baseball field, a 2-acre area south of the main buildings. This area requires further evaluation to characterize the radiological conditions. Remediation at the facility has been ongoing using criteria identified in the license. A comprehensive RP for addressing the remaining contaminated areas of the site is under development, but has not yet been submitted to NRC. In general, remediation work was scheduled so that the activities did not interfere with the manufacturing operations. FSSs for the remediated areas have been performed, but FSS documentation has not been transmitted to NRC.

## 3.0 Major Technical or Regulatory Issues

Although a significant amount of decontamination work has been performed in both interior and exterior areas of the facility, the licensee has not completed a FSS report. Monitoring well data for identification of any groundwater issues is limited. PADEP has interest in the site, particularly in the suspect baseball field area.

## 4.0 Estimated Date For Closure

12/01/2008

# Department of the Army (Ft. McClellan)

## 1.0 Site Identification

Location:              Fort McClellan, AL
License No.:           01-02861-05
Docket No.:            030-17584
License Status:        Active
Project Manager:       Orysia Masnyk Bailey

## 2.0 Site Status Summary

This site was licensed from 1956 until 1973 and from 1980 until the present. Building 3192 housed a hot cell for the fabrication of cesium and cobalt sources and as a result, building surfaces, soil and below ground tanks became contaminated. A license (01-02861-04) was issued for the possession of this residual contamination in Buildings 3182 and 3192 in 1974. Starting in 1980, the licensee performed closeout surveys for most of its authorized places of use under the Broad Scope license. Buildings 3182 and 3192 were remediated under a DP dated March 31, 1995. License No. 01-02861-04 was terminated on October 19, 1998, following an NRC confirmatory survey. Under the Broad Scope license, the Army performed closeout surveys and NRC performed confirmatory surveys of several areas including the burial areas at Iron Mountain and Rattlesnake Gulch, and of Buildings 1081, 2281, 3180, 3181, 3182, 3185, T-810, T-811, T812, T-836, T-837 and Alpha Field. Based on a characterization survey that caused the licensee to suspect the presence of discrete sources, a DP dated March 2, 1999, was submitted for the remediation of the burial mound at Pelham Range. The mound was found to include discrete sources (Co-60) and some Sr-90. The licensee provided a closeout survey. An NRC confirmatory survey was performed in July 2003. NRC and licensee survey results disclosed no contamination significantly above background levels. Except for the Pelham Range, the remainder of the Fort McClellan property has been turned over to the State of Alabama. The staff is not aware of a specific estimate for the cost of decommissioning.

## 3.0 Major Technical or Regulatory Issues

As part of the FSS, the Army performed a fly over survey of the base which resulted in the discovery of an additional burial ground. Elevated readings were seen at the Anniston City Recreational Area (LaGarde Park), a property that the Army donated to the city in 1976. Ground investigation disclosed the presence of soil contaminated with cesium and cobalt. USACE took responsibility for the site. In August 2003, USACE initiated clean up action at this site under CERCLA. Under this cleanup, contaminated soil was identified and removed as funds allowed. USACE returned to the site in 2004 to perform a characterization survey to gather data for a risk analysis regarding the Pelham Range burial area. The license will be terminated when USACE remediates the remaining burial area. The State of Alabama and the EPA are monitoring clean up activities.

## 4.0 Estimated Date For Closure

12/01/2006

# Eglin Air Force Base

## 1.0 Site Identification

| | |
|---|---|
| Location: | Walton County, FL |
| License No.: | 42-23539-01AF |
| Docket No.: | 030-28641 |
| License Status: | Active |
| Project Manager: | Robert Evans |

## 2.0 Site Status Summary

The Air Force submitted a DP to the NRC in May 2002, for a former depleted uranium munitions testing facility at Eglin Air Force Base. Supplemental information was provided to the NRC on November 1, 2002, August 21, 2003, October 27, 2004, and January 13, 2005. The NRC is considering the issuance of an amendment to Materials License 42-23539-01AF which will approve the DP. Test Area C-74L is located in Walton County, Florida, within the north-central portion of Eglin AFB. The test area currently consists of a 4-acre radiologically controlled area, fire control/ballistics building, gun corridor, target area, well house building, drum storage area, and surrounding land. On April 14, 2005, a draft environmental assessment (EA) was submitted to State of Florida for review. In addition, the FSS and confirmatory survey were completed during May 2005. The EA and FONSI were published in the Federal Register on July 11, 2005. The DP was approved by the NRC on January 20, 2006. The licensee is expected to submit the FSS report to the NRC during 2006.

From 1974 to 1978, the area was used for pre-production testing of a gun system which used depleted uranium ammunition. An estimated 16,315 pounds of depleted uranium were expended at the site.

## 3.0 Major Technical or Regulatory Issues

This site does not require a consultation with EPA because the post-remediation survey results are well below the criteria required for a consultation.

## 4.0 Estimated Date For Closure

12/06

# Engelhard Minerals - Great Lakes

## 1.0 Site Identification

| | |
|---|---|
| Location: | Great Lakes, IL |
| License No.: | SMC-01207, SUC-01332 |
| Docket No.: | 040-08306, 040-08680 |
| License Status: | Terminated |
| Project Manager: | Eugenio Bonano |

## 2.0 Site Status Summary

Engelhard Minerals & Chemicals Corporation (Engelhard), which is no longer in business, was licensed to repackage and ship monazite sand from the Great Lakes Naval Training Center to other AEC/NRC licensees. The area was used by the U.S. General Services Administration (GSA), which transferred control to the Defense Logistics Agency. The Engelhard license to ship the material was terminated in 1975. The former licensee was authorized to possess natural thorium (Monazite Sand). The Navy, which is the site owner, assumed responsibility for the Great Lakes site cleanup. A scoping survey conducted in March 2000, indicated radiological concentrations of Th-232 ranging from 0.93 pCi/g to 64.31 pCi/g with an average concentration of approximately 17.0 pCi/g. The monazite sand encompasses an area of approximately 90,000 square yards in a former tank farm area located within the boundaries of the Great Lakes Naval Training Center. Due to the relatively insoluble nature of the thorium, groundwater impact is not a concern. Characterization, cleanup, and FSSs are being done in three phases. In Phase 1, the site formerly known as Tank farm #5, was characterized. The entire tank farm was surveyed and surface soil samples were collected and analyzed for Thorium-232. The remainder of the Tank farm was fenced to restrict access pending further investigation and remediation. In Phase II, an excavated soil pile area was remediated and the contaminated soil was shipped for disposal. The third phase involves additional remediation and a FSS of the north fence area.

## 3.0 Major Technical or Regulatory Issues

This is an unlicensed facility. The Navy has assumed responsibility for the clean-up of this formerly licensed site. The Navy has been working cooperatively with the NRC and has agreed to employ NRC regulations and guidance documents, such as MARSSIM and NUREG -1757, to clean-up the site. On July 7, 2005, NRC inspectors identified additional thorium-232 contamination outside of the site boundaries east of the affected areas near a stream. The Navy has placed a cordon around this area. On July 12, 2005, NRC met with members of the Navy's Radiological Affairs Support Organization (RASO), and personnel from the Great Lakes Environmental Department to discuss the need to further characterize the site, establish new site boundaries, and develop new work plans (remediation and final status survey) for the site. Establishing a new site-specific Derived Concentration Guideline Level (DCGL) was discussed, including the submittal of new timelines to the NRC for the completion of the project. Based on the identification of the offsite contamination, the Navy will not be able to release the site for unrestricted use by December 2005, as was earlier planned. The Navy has stopped all decommissioning activities until further notice.

## 4.0 Estimated Date For Closure

TBD

# FMRI (Fansteel), Inc.

## 1.0 Site Identification

Location:           Muskogee, OK
License No.:        SMB-911
Docket No.:         040-07580
License Status:     Expired
Project Manager:    Jim Shepherd

## 2.0 Site Status Summary

The Muskogee site originally comprised about 52 hectares (110 acres) on the Arkansas River (Mile 395). It is about 4 kilometers (2.5 miles) from the center of the City of Muskogee, between the river on the east, Highway US-62 on the south, and the Muskogee Turnpike on the west. In 1996, 14 hectares (35 acres) know as the Northwest Property was released from the license.

The Muskogee facility, owned and operated by Fansteel Inc., produced tantalum and columbium metals from 1957 until it ceased operations in 1990. The raw materials used for tantalum and columbium production contained uranium and thorium as naturally occurring trace constituents. These radioactive species were present in the process raw materials at an approximate concentration of 0.1 percent uranium oxide and 0.25 percent thorium oxide. This concentration is sufficient to cause the ores and slags to be classified as source materials and issued a license by the AEC in 1967. The radioactive residues from the process were placed in several sludge ponds north of the process building. Other liquid waste went to several ponds in the southern part of the site.

Radioactive contaminants at the site include natural uranium, natural thorium, and decay products. Chemical contamination are also present in the form of metals including tantulum, niobium, chromium, antimony, tin, barium, arsenic; ammonia fluoride and methyl isobutyl ketone. In 1993, the licensee performed a characterization survey to determine existing conditions site wide. Radiological survey activities were conducted over the interior and exterior of the site structures and the open land areas of the site. Buildings and equipment associated with the ore-processing activities include the Chemical "C" Building, the Chemical "A" Building, and the R&D Building. The Chemical "C" Building is contaminated throughout by radioactive ore residues. Isolated areas of radioactive contamination were also identified in some of the other site buildings. Characterization surveys also identified the highest concentrations of radiological contaminants in Pond Nos. 2 and 3. Survey data indicate that the Th-232 and U-238 are present with their radioactive progeny in secular equilibrium. The U-235 decay series is also present, because U-235 constitutes 0.7 percent by weight (approximately 2.3 percent by radioactivity) of naturally occurring uranium.

NRC granted Fansteel a license amendment dated March 25, 1997, to complete the reprocessing of ore residues (WIP), calcium fluoride residues, and wastewater treatment residues containing uranium and thorium, in various site impoundments. Fansteel also planned to place the residue of these operations into an on-site disposal cell in accordance with 10 CFR 20.1403; this cell never received NRC approval.

In November, 2001, Fansteel suspended all operations at the Muskogee site, and in January, 2002, filed for bankruptcy protection under Chapter 11. Subsequently, NRC drew on the financial assurance instruments and that money is now in a standby trust. The license expired in September, 2002. A request for renewal was denied because the licensee stated it had ceased operations and intended to remediate the site for unrestricted use. Conditions of the license related to material control remain in effect in accordance with 10 CFR 40.42(c).

In July, 2003, Fansteel submitted: i) its DP; ii) a request for exemption from financial assurance requirements; and iii) a request for authorization to transfer the site license to a subsidiary to be formed as part of the bankruptcy reorganization plan. In this DP, the licensee revised the cost estimate for decommissioning to approximately $42 million from that in the bankruptcy filing of $57 million. On November 17, 2003, the bankruptcy court approved Fansteel's corporate reorganization plan to divide the company into two parts, with the second part going to the commercial creditors. FMRI Inc. (FMRI), a new subsidiary of Reorganized Fansteel, would become the licensee for the Muskogee site.

On December 4, 2003, NRC approved the DP, the request for exemption to financial assurance requirements, and the license transfer authorization, subject to the bankruptcy reorganization plan becoming effective. The approved DP outlines a phased approach to remedial activities that focuses on the most risk-significant areas and accomplishes those activities first. The approval also authorized FMRI to draw up to $2 million from the standby trust for remediation activities if it has insufficient funds from Fansteel to continue the work. This agreement was subsequently revised to authorize FMRI to draw additional monies from the fund for waste disposal as part of Phase 1 activities. The reorganization plan and NRC's approvals became effective on January 23, 2004.

Phase 1 of the DP states that the WIP in Ponds 2 and 3 will be removed from the site and sent to the White Mesa facility operated by International Uranium Corp. (IUC). Phase 1 was scheduled to commence in September, 2004. FMRI did not commence remediation activities until about June, 2005. In order for IUC to receive the material it must have a license amendment approved by the State of Utah. IUC submitted the application on April 8, 2005. On June 13, 2006, Utah issued the amendment authorizing receipt of FMRI material. In May, 2005 FMRI began a process of air drying and bagging the WIP in Pond 3 in preparation for shipment to IUC.

## 3.0 Major Technical or Regulatory Issues

Fansteel has provided a total of about $4.5 million in financial assurance. To date, FMRI has spent $2 million from the trust fund to assist in paying for the start of remediation activities. The original cost estimate for off-site disposal of all wastes greater than 10pCi/g total was $57 million. The revised cost estimate in the DP is about $30 million for solid waste, based on dose criteria of 10 CFR 20.1402 using an industrial land use scenario with no drinking water pathway. Fansteel estimated approximately $10 million additional for commitments for ground water remediation. Fansteel stated it is not able to provide additional financial assurance because of the bankruptcy proceeding. Instead, it signed unsecured promissory notes for the estimated costs. As of May, 2006, FMRI has made four withdrawals from the Trust, for a total of about $2.2 million, and one deposit from an insurance settlement of about $764 thousand. The remaining value of the fund is about $3 million.

FMRI did not commence remediation by September 1, 2004, as required by license condition, but did commence excavation of Pond 3 in June, 2005. On April 13, 2005, NRC issued a Notice of Violation (NOV) (EA-04-188) for failure to commence remediation as required by Condition 26 of SMB-911. NRC determined not to pursue the apparent violation, but to focus on FMRI meeting its completion date.

FMRI did not provide updates to annual financial projections (Table 15-12 of the DP) as required by its license. On July 26, 2005, NRC issued a NOV for failure to submit information as required by its license. FMRI responded that Fansteel, its parent, did not provide the information (FMRI is not an operating company and has no other revenue source). NRC did not consider FMRI's responses to be adequate and FRMI agreed to submit a request for license amendment to resolve the issue. NRC rejected FMRI's request because it did not meet the intent of the original condition nor NRC's information needs. FMRI requested a meeting with NRC to discuss a course of action to resolve the on-going violation.

In June, 2006 FMRI's excavation contractor stopped work in Pond 3 because of perceived difficulties in excavation near the center of the pond. Also, the time limit on the existing transportation contract expired before authorization to ship to IUC was granted; FMRI is negotiating a new contract. FMRI has supersacks sitting on the ground per a temporary exemption to a license condition specifying storage conditions. Because it cannot ship the material before the exemption expires in September, 2006, FMRI must request further licensing action on this matter.

There is high public interest from the State of Oklahoma, the Cherokee Nation, and the Port of Muskogee.

## 4.0 Estimated Date For Closure

12/12/2023

# Homer Laughlin

## 1.0 Site Identification

| | |
|---|---|
| Location: | Newell, WV |
| License No.: | SUB-00081 |
| Docket No.: | 040-01957 |
| License Status: | Terminated |
| Project Manager: | John Nicholson |

## 2.0 Site Status Summary

The Homer Laughlin China Company (Homer Laughlin) is a 37-Acre facility located on the banks of the Ohio River, in Newell, West Virginia. The town is located in West Virginia's Northern Panhandle, approximately 40 miles Northwest of Pittsburgh, PA. The facility is an active business, manufacturing retail and commercial dinnerware. It encompasses several buildings on 37 acres and employs over 1100 individuals.

Homer Laughlin is a formerly-licensed site. Homer Laughlin was licensed by the AEC for possession of 100,000 pounds of source material used as a glazing agent (up to 20% uranium) in the production of ceramic tableware. The license was terminated in 1972 based upon a letter from Homer Laughlin stating that all remaining licensed materials had been returned to their supplier. A review of the terminated license file determined that no record of licensee closeout survey or NRC confirmatory survey was performed. In 1994, approximately 500 pounds of depleted uranium oxide (U3O8) sand was discovered on the property. A contractor was hired to survey areas where licensed materials were used and stored and provide a radiological characterization of material in the facility. Several areas of fixed and removable contamination exceeding NRC limits for unrestricted use were identified during the characterization survey. NRC issued a Confirmatory Action Letter to HLC requiring a commitment to package and dispose of the bulk source material, limit access to contaminated areas, and submit a DP. After NRC approved the DP in January 1995, Homer Laughlin initiated facility decommissioning. Homer Laughlin did not complete decommissioning in production areas because they were unable to remove fixed contamination from surfaces of equipment and structures which exceeded NRC unrestricted release guidelines using conventional techniques. At various times during the period 1996-2004, Homer Laughlin provided additional information to NRC to refine their computer-based risk analysis, to demonstrate that the facility meets the 25 mrem/yr unrestricted release limit of the LTR. In March, 2005, NRC accepted Homer Laughlin's revised risk assessment, pending the verification of certain assumptions made in the analysis (contaminated surface area, no contamination above background on kiln floor, secular equilibrium of U-238, no other alpha emitters considered). Although the facility likely meets the LTR, the waste material from decontamination activities remains on site, packaged and in storage.

## 3.0 Major Technical or Regulatory Issues

Waste material from site decommissioning activities has been packaged and is being stored in a posted, and infrequently-used area of the plant. Homer Laughlin is currently performing characterization of the material. An initial random sampling characterization was conducted in the first quarter of 2006, and a complete characterization is expected to be completed by the end of 2006. Cost estimates for waste disposal will be solicited from waste vendors when characterization is completed.

## 4.0 Estimated Date For Closure

TBD

# Jefferson Proving Ground

## 1.0 Site Identification

| | |
|---|---|
| Location: | Rock Island, IN |
| License No.: | SUB-1435 |
| Docket No.: | 040-08838 |
| License Status: | Possession-only |
| Project Manager: | Tom McLaughlin |

## 2.0 Site Status Summary

Contamination on site consists of depleted uranium (DU) in the soil. Additionally, there is a concern for future groundwater contamination. The site has been closed for the testing of all ordnance including DU rounds since 1995. The monitoring of DU in soil, groundwater, surface water, and sediment continues on a semi-annual basis. The U.S. Army submitted a revised DP in June 2002. The Army has submitted a request for an alternate decommissioning schedule in order to collect data needed for decommissioning the site under restricted release. The Army wants to keep its possession-only license for a 5-year period at which time it will submit a revised DP. There are no immediate radiological hazards at the site. Unexploded ordnance (UXO) at the site represents a significant non-radiological hazard. The staff does not have an estimate of the cost of decommissioning.

The site has been closed for the testing of all ordnance including depleted uranium rounds since 1995. The license was amended on May 8, 1996, resulting in the area south of the firing line being released for unrestricted use. The area north of the firing line contains about 70,000 kg of DU along with a large amount of UXO. The Army submitted a revised DP and an Environmental Report on June 27, 2002, which were accepted for technical review on October 1, 2002. During a limited technical review, the NRC staff concluded that site-specific data were needed in order to validate any off-site transport model. On February 4, 2003, the U.S. Army submitted a letter to NRC requesting an alternate schedule under 10 CFR 40.42(g)(2) that would create a 5-year renewable possession-only license for an indefinite time period. Subsequently, the Army has withdrawn its request for a 5-year renewable possession-only license for an indefinite time period and is collecting data leading towards decommissioning.

## 3.0 Major Technical or Regulatory Issues

The presence of UXO, the associated risk, and cost for cleanup of this material, as well as potential contamination of groundwater, are complicating remediation. The Army has signed a memorandum of agreement with the Department of the Interior and the Department of Defense (Air Force) for long-term institutional control of the site. In January 2000, Safe the Valley, a local environmental group, requested a hearing on the DP, citing that the DP does not adequately describe the decommissioning process and does not provide adequate assurance for long-term control. The hearing has been extended to include the proposed amendment. No financial assurance issues have been identified at this time.

## 4.0 Estimated Date For Closure

09/30/2010

# Kaiser Aluminum

## 1.0 Site Identification

Location:          Tulsa, OK
License No.:       STB-472
Docket No.:        40-02377
License Status:    Terminated
Project Manager:   John Buckley

## 2.0 Site Status Summary

The Kaiser Aluminum and Chemical Corporation (Kaiser) facility is located at 7311 East 41st Street in Tulsa, OK. The known affected area covers approximately 9 acres.

On March 7, 1958, NRC issued Source Material License No. C-4012 to Standard Magnesium Corporation (Standard Magnesium), a Division of Kaiser Chemical Company, for possession of magnesium-thorium alloy. Standard Magnesium purchased magnesium-thorium scrap metal for reclaiming purposes. The end product from Standard Magnesium's manufacturing process was magnesium anodes used for cathodic protection on items such as tanks and pipelines. NRC License No. STB-472 superceded License No. C-4012 on November 22, 1961. On June 5, 1968, License No. STB-472 was amended to include the possession of uranium. There is no evidence to indicate that uranium was ever received or processed on site. On March 16, 1971, License STB-472 was terminated. On November 17, 1993, NRC surveyed the Kaiser facility to assess the potential for residual contamination at the site. Contamination was found on the surface, indicating that waste magnesium-thorium slag was disposed of in the past. Off-site residual thorium contamination was first identified in June 1994. The off-site thorium contamination is due to slag dumping in areas to the east and south of the current Kaiser boundary, on property which belonged to Standard Magnesium during licensed operations. The NRC added Kaiser to the SDMP on August 19, 1994. Kaiser began decommissioning the site after approval of its DP in June 2003, and completed decommissioning in September 2006. It is estimated that NRC will approve Kaiser's remediation activities at the Tulsa site in October 2006.

## 3.0 Major Technical or Regulatory Issues

To date there is minimal public interest in the decommissioning activities at the site. The staff has not identified any major off-site environmental issues that will no be addressed during remediation of the facility.

## 4.0 Estimated Date For Closure

10/30/2006

# Kerr-McGee - Cimarron

## 1.0 Site Identification

Location:            Oklahoma City, OK
License No.:         SNM-928
Docket No.:          70-0925
License Status:      Active
Project Manager:     Ken Kalman

## 2.0 Site Status Summary

The 840 acre Cimarron site in Crescent, Oklahoma is situated along the southern bank of the Cimarron River approximately 30 miles north of Oklahoma City. Most of the site has been decommissioned and released for unrestricted use. Uranium contamination in excess of release criteria is in the groundwater at Burial Area 1 and around Well 1319. Technetium (Tc)-99 exceeding release criteria is in the groundwater in the vicinity of Waste Pond 1 and 2. Concentrations of Tc-99 within applicable release criteria have also been found in Burial Area 1. The site is also licensed for onsite disposal of up to 500,000 cubic feet of Option 2 [of the 1981 Branch Technical Position (BTP)] contaminated soil in Subarea N. NRC staff reviewed Cimarron's Subarea N Report (submitted in January 2002) and performed its independent confirmatory survey in June 2002. Due to a recent occurrence of groundwater exceeding the 180 pCi/l release limit in a nearby portion of Subarea K, NRC is delaying release of Subarea N until the groundwater issue is resolved. There are no immediate radiological hazards at the site. The licensee estimates the cost of decommissioning to be approximately $3.6 million. No financial assurance issues have been identified at this time.

The Kerr-McGee Corporation (KMC) operated two plants at the Cimarron facility between 1965 and 1975, each under its own separate AEC license. License SNM-928 was issued under 10 CFR Part 70 for the uranium fuel fabrication facility, and License SNM-1174 was issued for the mixed oxide fuel fabrication (MOFF) facility. Subsequently, in 1988, Cimarron Corporation (Cimarron), a wholly-owned subsidiary of KMC, became responsible for the Cimarron facility. NRC terminated SNM-1174 by letter dated February 5, 1993. Although License SNM-1174 was terminated, the MOFF plant building exterior surfaces and grounds were retained under License SNM-928. Cimarron began decommissioning in 1977 and has completed most of the decommissioning activities needed for NRC to release the Cimarron site for unrestricted use and to terminate License SNM-928. The primary remaining activity to be completed is groundwater remediation. Cimarron is considering several alternatives for groundwater remediation including natural attenuation, excavation, and the use of institutional controls. The final choice will be dependent on coordination among Cimarron, Oklahoma Department of Environmental Quality, and NRC. Cimarron anticipates submitting its proposal to NRC in October 2006. In 2006, Kerr McGee Chemicals Worldwide LLC was changed to Tronox Worldwide LLC, a wholly owned subsidiary of Tronox Inc. Cimarron Corporation is still the licensee for the site.

## 3.0 Major Technical or Regulatory Issues

One significant regulatory issue that was resolved was whether NRC will allow Cimarron to remediate the groundwater under SDMP criteria or if Cimarron's choice of remediation technique will require the use of criteria from the LTR. NRC determined that Cimarron's preferred technologies of "pump and treat" or excavation could be implemented using the SDMP criteria. Depending on the groundwater remediation technique that Cimarron proposes, ODEQ may raise concerns regarding the disposal of soils or effluent discharges.

## 4.0 Estimated Date For Closure

05/01/2010

# Mallinckrodt Chemical Inc.

## 1.0 Site Identification

Location:            St, Louis, MO
License No.:         STB-401
Docket No.:          40-6563
License Status:      Possession-Only
Project Manager:     Amir Kouhestani

## 2.0 Site Status Summary

Contaminants at the Mallinckrodt Chemical, Inc., (Mallinckrodt) site are: U-238; U-235; U-234 and progeny; Th-230; Ra-226; Th-232; Th-228 and progeny; Ra-228; and K-40. Although total uranium was detected in the filtered samples at elevated concentration, it was concluded that these detections do not present a groundwater ingestion hazard since the perched groundwater in the upper zone is not a drinking water source. Decommissioning at the Mallinckrodt site will take place in two phases. Phase 1 addresses the decommissioning of the buildings and equipment to the extent that whatever remains on site will be released for unrestricted use. Phase 1 was completed in December 2004. Phase II will complete the decommissioning of the building slabs and foundations, paved surfaces, and all subsurface license related materials to the extent that they can be released for unrestricted use. Mallinckrodt submitted the Phase 1 DP on November 20, 1997. NRC approved the Phase 1 DP on May 3, 2002. Remediation at the site began in July 2002. Mallinckrodt submitted its Phase II DP on May 15, 2003. The staff is now reviewing the DP. Mallinckrodt is requesting to remediate the site to meet the unrestricted release criteria of 10 CFR Part 20, Subpart E. The estimated cost of decommissioning as presented in the Phase II DP is approximately $21 million.

Mallinckrodt has been operating at the St. Louis Plant since 1867 producing various products including metallic oxides and salts, ammonia, and organic chemicals. From 1942 to 1957, Mallinckrodt was under contract with the Manhattan Engineering Project and the Atomic Energy Commission (MED-AEC) to process uranium ore to produce uranium for development of atomic weapons. The St. Louis Plant, comprised of over 50 buildings on approximately 43 acres, is subdivided into smaller areas, called plants, based on the similarity of operations being performed. In 1961, Mallinckrodt was issued License STB-401 to extract columbium and tantalum (C-T) from natural ores and tin slags. From 1961 to 1974, Mallinckrodt purchased feed materials for C-T processing. Processing occurred from 1975 to 1985. The ores and processing byproduct materials contained uranium and thorium isotopes. C-T processing was shutdown from 1985 through early 1987, when Mallinckrodt began a two month pilot production run. During the pilot production run approximately 20,000 pounds of tin slag were processed. In July 1993, NRC amended Mallinckrodt's license to a possession only license for decommissioning and license termination. Approximately 6 Ci of natural uranium and 19 Ci of natural thorium isotopes were contained in the ores and tin slags processed under License STB-401.

Radiological contamination at the site resulted from MED-AEC and C-T processing activities. MED-AEC contamination is being removed by the USACE under FUSRAP. USACE developed a preferred cleanup method for the MED-AEC contamination, based on the data and findings presented in five documents: (1) Remedial Investigation Report; (2) Baseline Risk Assessment;

(3) Initial Screening of Alternatives; and (4) Feasibility Study & Proposed Plan, and (5) Record of Decision. There are no financial assurance issues identified at this time. Public interest in the site is high, however, concern over decommissioning activities is low. The staff has not identified any major off-site environmental issues that will not be addressed during remediation of the facility.

## 3.0 Major Technical or Regulatory Issues

Remediation of MED-AEC radiological constituents is currently being performed under the FUSRAP by USACE. USACE and Mallinckrodt have yet to agree on who has remediation responsibility for several areas within the facility. Applicable license site boundaries with respect to NRC licensed material and FUSRAP material has been examined. Staff, in reviewing Phase II DP, is also assessing any impact that extent of site boundaries determination may have on facility's dose modeling. The State of Missouri is concerned about chemical contamination on site. The State, under its EPA delegated authority, has issued Mallinckrodt a RCRA permit for their eventual site chemical cleanup. Boundary issues, subsurface contamination responsibilities, and post remediation groundwater monitoring issues have been identified by these parties. Staff is in the process of evaluating and bringing to closure licensee's responses to staff RAIs for the Phase II DP.

## 4.0 Estimated Date For Closure

07/01/2008

# Molycorp

## 1.0 Site Identification

Location:           Washington, PA
License No.:        SMB-1393
Docket No.:         040-08778
License Status:     Possession-only
Project Manager:    Tom McLaughlin

## 2.0 Site Status Summary

This site is located 56.3 Km (35 mi) southwest of the City of Pittsburgh in Canton Township, less than 0.8 Km (0.5 mi) southwest of the City of Washington, PA. Molycorp produced a ferroniobium alloy from an ore that contained natural thorium with some uranium. The operation resulted in the production of thorium-bearing slag that was used as fill over portions of the site. Average thorium concentrations over most of the site are between 100 and 200 pCi/g. In some locations, the contamination extends up to 3 m (10 ft) in the subsurface soil. Estimates of total waste volumes range from 45,846 - 114,615 m3 (60,000 - 150,000 yd3). Molycorp submitted its original DP in July 1995. The DP proposed on-site storage, followed by permanent disposal of the waste, from both the Washington and York sites, in an impoundment on the Washington site. Because on site disposal would have exceeded the SDMP Action Plan criteria, the NRC staff requested that Molycorp submit an environmental report (ER) as part of the DP. The licensee supplemented the 1995 DP with an ER in April 1997. After consultation with NRC staff, the licensee stated its intention to submit a revised DP in two parts: Part I of the DP would address cleanup of the contaminated portion of the site and comply with the SDMP criteria; and Part II would address disposal of material from the York and Washington sites in an impoundment on the Washington site and would comply with the LTR. NRC staff agreed to this approach and a revised DP (Part I) was submitted on June 30, 1999. The staff approved the Part I DP on August 8, 2000. In January 2001, Molycorp withdrew its amendment request for approval of the Part II DP (on site disposal cell). While Molycorp will continue to decommission the Washington facility under its previously approved Part I DP, it will now dispose of the material off site and will ultimately seek a unrestricted release of the site. On February 26, 2001, Molycorp informed NRC that it finished removal of all its stored above ground waste and shipped the material to the Envirocare facility in Clive, Utah. Molycorp now has torn down all of its buildings and has sent non-radioactive contaminated materials off-site and radioactive materials to Waste Control Specialists (WCS). All buildings and foundations have been removed from the site. The licensee has conducted a new site characterization to determine the amount and extent of contamination and a path forward for decommissioning the surface and subsurface soils. The license estimates the cost of decommissioning to be approximately $30.3 million.

Public concern in the Canton Township, City of Washington area, is low. Congressional interest also mirrors that found in the local community. The NRC has conducted two local public meetings to keep interested parties informed, the second of which was attended by over 300 people. On March 20, 2001, DWM staff participated in an "open house" style public meeting in Washington, PA, hosted by the Agency for Toxic Substances & Disease Registry (ATSDR). Other agencies participating included PADEP and the Pennsylvania Department of

Health. The Commonwealth of Pennsylvania may become the regulatory authority for this site before the completion of the decommissioning.

## 3.0 Major Technical or Regulatory Issues

Public concern in the Canton Township, City of Washington area, is low. Congressional interest also mirrors that found in the local communities.

## 4.0 Estimated Date For Closure

06/01/08

# NWI Breckenridge

## 1.0 Site Identification

Location:           Breckenridge, MI
License No.:        SMB-0833
Docket No.:         040-06264
License Status:     Terminated
Project Manager:    Peter J. Lee

## 2.0 Site Status Summary

Between 1967 and 1970, Michigan Chemical Corporation (MCC) managed the site and used it for the disposal of process wastes from a yttrium recovery operation. These disposal activities were authorized under AEC License SMB-0833, and were performed in accordance with 10 CFR 20.304. The buried waste material is a solid waste byproduct, known as filtercake, which originated from a rare-earth metal (yttrium) extraction process. Disposal records reported that the filtercake was typically a dense, clay-like material that contained elevated levels of naturally occurring uranium and thorium. After site operations ceased, AEC License SMB- 0833 was terminated. In addition to the buried wastes, thorium and uranium contaminated surface and subsurface soil has been identified at several locations in open land areas on the site. Several radiological evaluations have been performed in recent years. The most recent of these evaluations took place in November 2001, and led to completion of a characterization report which was submitted to the NRC in March, 2002. The average concentrations were determined to be about 240 picocuries per gram (pCi/g) of thorium-232 and 150 pCi/g of uranium-238.

The site is currently being remediated in accordance with a bankruptcy agreement, by the contractor Environ International Corporation (ENVIRON). SCIENTECH has worked on the remedial project since 1997, and has been retained as ENVIRON's subcontractor. ENVIRON submitted a RP in March 2004. NRC approved the RP in August 2004.

## 3.0 Major Technical or Regulatory Issues

Based on the remedial work plan prepared by SCIENTECH, there are seven confirmed waste areas (CWA) and two potential waste areas on the site. In October 2004, the SCIENTECH completed excavation of waste at CWA-2 and CWA-7. During the excavation of CWA-2 and CWA-7, the cover soils were found contaminated and additional veins of the waste were discovered. As a result of the unexpected increase in the waste volume, the Custodial Trust has insufficient funds to complete the remediation at this time. ENVIRON shut down the operation and transported all the excavated waste material for offsite disposal on November 19, 2004. ENVIRON has basically spent $0.75 million previously approved by the Bankruptcy Settlement Agreement.

To request additional funds to complete remediation of the site, ENVIRON is required to determine the volume of the waste remaining. ENVIRON agreed to prepare a work plan that will detail the procedures to be followed for the further characterizing the site and reevaluation of the derived concentration guideline levels for the site.

ENVIRON submitted the revised dose assessment on April 25, 2006 and supplemental site characterization plan on September 19, 2006, to quantify the accurate waste volume. The plan is currently under review by the staff.

## 4.0 Estimated Date For Closure

TBD

# Pathfinder

## 1.0 Site Identification

Location:          Sioux Falls, SD
License No.:       22-08799-02
Docket No.:        030-05004
License Status:    Possession-only
Project Manager:   Chad Glenn

## 2.0 Site Status Summary

The Pathfinder Atomic plant was designed to generate 66 MW of electric energy and operated from August, 1966 to September, 1967. The nuclear fuel was shipped off-site in 1970 and the plant was placed in SAFSTOR in 1971. In September 1972, Pathfinder's 10 CFR Part 50 operating license was terminated and the current 10 CFR Part 30 byproduct license was issued. The reactor and fuel storage facilities were decommissioned in 1991 under Reg. Guide 1.86. In November 1992, NRC amended the license to authorize the unrestricted release of the reactor building, fuel storage building, and waste storage building; to demolish the reactor building; and to authorize the possession of fixed activation products at the Pathfinder site. During its brief operating period, a relatively small amount of radioactive contamination was found in the steam turbine and auxiliaries. These systems are collectively referred to as the Balance of Plant (BOP) systems to distinguish it from primary power plant systems such as the reactor and its auxiliaries. The BOP was later decontaminated and disconnected from the reactor plant steam source. The BOP did not receive any additional radioactivity from any source after this period. The residual radioactivity contained within the BOP is a byproduct of materials activated during operation. The BOP was then integrated into a fossil-fueled peaking plant with gas/oil package boilers supplying steam to operate the existing turbine. The Pathfinder plant that utilized the original nuclear plant's BOP continued to operate on peaking duty until July 13, 2000, when the cooling tower collapsed in a storm. For economic reasons, the decision was made to cease operations of the peaking plant. In February 2003, Xcel Energy notified NRC that it had permanently ceased operating activities at the Pathfinder generating plant. In February 2004, Xcel Energy submitted a DP and license amendment request to authorize decommissioning activities at Pathfinder. On May 27, 2005, NRC approved the DP. The removal of the radioactive byproduct material within the steam, feed-water, and condensate portions of the BOP is the subject of the Pathfinder decommissioning. According to the licensee, the contamination consists of Co-60 (40 millicuries) and Zn-65 (1 millicurie). This material is in the form of fixed activation products in the BOP. Xcel Energy will remediate the contaminated areas to permit unrestricted use of the Pathfinder site.

## 3.0 Major Technical or Regulatory Issues

None.

## 4.0 Estimated Date For Closure

05/27/2007

# Quehanna (Formerly Permagrain Products, Inc.)

## 1.0 Site Identification

| | |
|---|---|
| Location: | Harrisburg, PA |
| License No.: | 37-17860-02 |
| Docket No.: | 030-29288 |
| License Status: | Possession-only |
| Project Manager: | James Kottan |

## 2.0 Site Status Summary

The Commonwealth of Pennsylvania owns the site and had leased it to Permagrain Products, Inc. (PPI) for the operation of a Co-60 irradiator. After PPI declared bankruptcy in 2002, the license was transferred to the Commonwealth of Pennsylvania. Sr-90 is the main contaminant of concern at the facility, and was used in the manufacture of thermoelectric generators. Sr-90 contamination is found in buildings as well as in surface and subsurface soil. Contaminated groundwater is not present at the site. The decommissioning, which started in July 1998 is being performed by Energy Solutions (formerly Scientech). Areas which do not meet NRC criteria for unrestricted use were identified as the six hot cells, their respective isolation rooms, two ventilation systems, an overhead crane system, a number of ancillary rooms, and the wastewater treatment building. Decontamination and demolition of the cell structures was completed in 2004. Decontamination of the service area floor is complete. FSS was initiated and was completed in December 2004. The licensee's FSS Report along with a request to terminate the license was submitted in February 2005. An NRC confirmatory survey of the facility was conducted in May 2005. The confirmatory survey identified numerous areas of contamination in excess of the NRC approved limits for unrestricted release. Based on these survey results, the licensee performed additional surveys and concluded that some type of migration of radioactive material is taking place in the concrete at the facility. On March 13, 2006 NRC received a revised DP from PADEP proposing to complete site closure under 10CFR20, Subpart E. The total cost of decommissioning to date has been approximately $25 million. The licensee expects that an additional $2 to 3 million will be needed to complete decommissioning.

## 3.0 Major Technical or Regulatory Issues

None.

## 4.0 Estimated Date For Closure

05/01/2007

# Royersford Wastewater Treatment Facility

## 1.0 Site Identification

Location:            Royersford, PA
License No.:         Non-Licensee
Docket No.:          NA
License Status:      Unlicensed
Project Manager:     Betsy Ullrich

## 2.0 Site Status Summary

The Royersford Wastewater Treatment Facility (RWTF) receives waste water that contains radionuclides from wastewater generated by a nuclear laundry, UniTech Services Group (UniTech), formerly known as Interstate Nuclear Services (INS). These discharges began in the late 1980's. Elevated levels of radioactivity and radiation have been detected at the RWTF since 1986, in the secondary digestor sludge and the resulting solid products from the dewatering of the secondary digestor sludge. The main contaminants are Co-60 and Cs-137. Unitech has been in compliance with NRC regulations for disposal to the sanitary sewerage system with typical concentrations of radionuclides in the UniTech wastewater of less than 10% of the regulatory limits for disposal to the sanitary sewer. The total amount of radionuclides released each year, other than tritium and carbon-14, ranges from 66 mCi to 492 mCi. In 2003, UniTech completed installing a pipe from their facility to the Schuylkill River, and obtained a National Pollutants Discharge Elimination System permit for discharge. RWTF has finished cleaning out the lines from UniTech, and cleaned the settling tanks, primary digestor, secondary digestor, etc. The only remaining radioactivity is in the reedbeds. RWTF is looking into options for the disposal of the reedbed sludge. To date, no estimate for the cost of decommissioning has been developed.

## 3.0 Major Technical or Regulatory Issues

The RWTF secondary digester sludge is a liquid containing 3%-6% solids. Samples of secondary digestor sludge have Co-60 concentrations typically in the range of 9,000-60,000 picocuries per liter (pCi/l), although individual samples have contained as much as 115,000 pCi/l. Cs-137 concentrations are in the range of 1,500-5,000 pCi/l. The disposition of the secondary digestor sludge by mechanical dewatering is performed once or twice each year. The resulting filtercake contains about 20% solids and has been disposed of at a municipal waste landfill. Filtercake samples contain in the range of 22-950 pCi/g for Co-60 and 8-112 pCi/g for Cs-137. Radiation levels measured typically are 80-100 microR/hr near contact with the filtercake. In 1990, the RWTF began using an onsite reedbed for biological dewatering of secondary digestor sludge. Resulting reedbed sludge is located on site in a 6-foot-high walled reedbed, with the height of the sludge rising as additional material is added. Reedbed sludge is a marsh-like material, containing up to 40% solids. Reedbed sludge samples contained from 77-950 pCi/g of Co-60 and 20-90 pCi/g of Cs-137. Radiation levels near the surface of the onsite reedbed have increased over time to the range of 800-1000 microR/hr. The reedbed reached its capacity in 2003, and the dried sludge needs to be removed and disposed. Two main issues are anticipated for closure of the reedbeds: (a) potential radiation doses to workers involved in removal of the reedbed sludge; and (b) disposal of the reedbed sludge. Potential dose can be estimated using the licensee's pathway analysis assumption that removal of the sludge would require 10 working days and the average dose rate in the reedbeds. During the year 2000, the average dose rate was 0.345 millirem /hr in the reedbeds; therefore, a person working in the reedbeds for 80 hours could receive 28 millirem, from external sources, during sludge removal. (The inhalation pathway is a factor of approximately 10,000 smaller, based on the INS pathway analysis.) Disposal of the reedbed sludge may be more complicated than past disposals of filtercake from mechanical dewatering of sludge to the municipal landfill. In 2000, the Commonwealth of Pennsylvania passed legislation that requires all landfills to have radiation monitors to survey incoming material, to ensure that no radioactive materials are disposed of in the State. The radiation levels from the reedbed sludge will likely be detected by such monitors. The radiation levels from filtercake may also be detected by radiation monitors, but no such disposals have been made since the legislation was passed. Disposal of dried sludge may not meet the requirements of low level waste sites, if the sludge contains substances that are considered hazardous materials (this is likely). It is not known at this time if the Commonwealth of Pennsylvania will continue to allow sludge from the RWTF to be transferred to a municipal landfill. NRC conducted field sampling at the site on January 10, 2006. Sample results were shared with the facility and with PADEP. NRC is considering a request by PADEP to complete a dose assessment for the reed-bed material. Public interest in the RWTF contaminated sludge is sporadic and is usually associated with issues at the landfill receiving the sludge.

## 4.0 Estimated Date For Closure

TBD

# Safety Light Corporation

## 1.0 Site Identification

Location:              Bloomsburg, PA
License No.:           37-00030-02; 37-00030-08
Docket No.:            030-05980; 030-05982
License Status:        Active
Project Manager:       Robert Prince

## 2.0 Site Status Summary

Safety Light Corporation (SLC) is an active site licensed to manufacture tritium exit signs, and for to perform site characterization and decommissioning activities. Contamination at the site is from the manufacturing operations of self-luminous watch and instrument dials and other items involving Ra-226, Cs-137, Sr-90, and Am-241. Radioactive waste was disposed on-site in three primary locations: silos, lagoons, and a waste dump. Primary soil contaminates include Ra-226 and Cs-137 with small amounts of Am-241. The onsite ground water is also contaminated with H-3, Sr-90, and Cs-137. In October and December 2000, SLC submitted a DP to NRC which called for a "task by task" approach to decommissioning because of limited funding availability. The DP presents decommissioning activities which will make the site suitable for unrestricted release. This approach was approved by NRC in December 2001, and on August 15, 2002, NRC amended the SLC license to approve the work plan for processing and sorting waste that was removed from two underground silos in the fall of 1999. NRC staff continues to coordinate activities with EPA and PADEP regarding remediation of the SLC site. An EPA Administrative Order of Consent with SLC for the sorting, characterization, and re-packaging of the drums of mixed waste and radioactive waste that were removed from the onsite silos, became effective on February 3, 2003. Under the EPA Emergency Removal Action, three shipments of radioactive material to an offsite disposal facility were completed by November 15, 2004. Disposal costs are expected to exceed the licensee's decommissioning funds, so EPA is expected to propose a unilateral Order and use EPA emergency removal funds to complete disposal of the underground silo waste. On September 23, 2004, EPA proposed adding SLC to the National Priority List (NPL). SLC was added to the NPL in an April 27, 2005 rulemaking (70 FR 21644). Remedial investigation studies for groundwater, buildings and soil were initiated in October 2004, January 2006 and May 2006, respectively. EPA has initiated preliminary work activities at the SLC site. These activities include preparations to dismantle various buildings and efforts to ship offsite the remaining radioactive waste containers that consist of previously packaged silo waste. The licensee estimates the cost of decommissioning to be approximately $29 million. An NRC analysis of the licensee's Decommissioning Cost Estimate concluded that the decommissioning cost for unrestricted release of the site by the licensee was estimated to be between $94 and $120 million and to be $50 million to $78 million for restricted release.

## 3.0 Major Technical or Regulatory Issues

Lack of financial assurance remains the key issue; effective remediation work cannot be performed because of limited funding. The licensee submitted a request for license renewal, which was received on April 29, 2004. On December 10, 2004, the application to renew SLC licenses 37-00030-02 and 37-00030-08 was denied by the NRC. In addition, an Order was issued instructing SLC to initiate procedures to terminate their licenses pursuant to 10 CFR

30.36. On January 13, 2005, the Atomic Safety Licensing Board (ASLB) heard a motion by SLC to set aside the immediate effectiveness of the Order suspending the license. A settlement agreement with NRC, SLC and Pennsylvania PADEP was approved by the ASLB on June 29, 2005. The ASLB decision became a final Agency decision on August 8, 2005. The renewed license requires SLC to develop a plan for the orderly shutdown of licensed activities and make prescribed monthly payments into the decommissioning trust fund during the license renewal period. The renewed license will expire on December 31, 2007.

## 4.0 Estimated Date For Closure

12/31/2007

# Salmon River

## 1.0 Site Identification

| | |
|---|---|
| Location: | North Fork, ID |
| License No.: | R-0230 and P-4001 |
| Docket No.: | 040-03400 |
| License Status: | Terminated |
| Project Manager: | Rafael Rodriguez |

## 2.0 Site Status Summary

The former Salmon River Uranium Development (SRUD) mill site consists of a 21.5-acre privately-owned land surrounded by United States Department of Agriculture Forest Service lands. Located along the Salmon River, 5 miles west of North Fork, ID, the site includes an abandoned mine, a large structure previously used for milling and chemical operations, and a tailings pond. The site was licensed from 1958 through 1959 by the AEC. Although both uranium and thorium ore were mechanically and chemically processed at the site, it is suspected that operations with source material at SRUD were very minor and only experimental in nature. There is also hazardous contamination in the mill (primarily, sulfuric acid) resulting from the operations at the facility between 1978-1979. Contamination is found inside the buildings, the tailings pond, and is believed to extend to an area of 2 acres around the mill. The site was identified as part of the terminated license review project conducted by the Oak Ridge National Laboratory (ORNL) during the 1990s. On May 2001, NRC visited the former SRUD site and identified thorium contamination in the form of partially processed ore. Laboratory results confirmed that the material onsite was "source material" (i.e., >0.05 wt% Th). On July 2001, the property owners were notified regarding the results of the site inspection. During 2004 and 2005, staff worked with the site owners, Idaho Department of Environmental Quality (IDEQ), and EPA to establish a path forward to address remediation of the site. On December 2005, NRC requested EPA's assistance in the remediation of the property, in part because of the existing hazardous contamination at the site. NRC and EPA conducted a site visit in June 2006, to assess the existing radiological and non-radiological contamination at the site. Currently, EPA is analyzing the samples and data collected during the site visit. A report documenting the results and stating whether the site warrants action by EPA is expected in 2006.

## 3.0 Major Technical or Regulatory Issues

Financial assurance and the existence of source material and hazardous waste on the property are the key issues. The existence of hazardous waste has triggered coordination with EPA. In August 2005, the site owner indicated that he is considering living on the property at some future time, and that he does not have the resources to decommission the site. Based on these facts staff options are very limited, in part due to the remote location of the property. NRC staff will continue working with EPA and IDEQ to address remediation activities.

## 4.0 Estimated Date For Closure

05/12/2012

# S.C. Holdings

## 1.0 Site Identification

Location:          Kawkawlin, MI
License No.:      SUC-1565
Docket No.:      40-09022
License Status:   Active
Project Manager:  David Nelson

## 2.0 Site Status Summary

The S.C. Holdings, Inc. (S.C. Holdings) site in Bay County, Michigan, is part of the former Hartley & Hartley Landfill. The site covers about 235 acres and part of the site is contaminated with thorium. The contamination came from magnesium-thorium alloy production at a defunct former licensee. The contaminated soil is covered with a clay cap and encapsulated with slurry walls and in two small piles covered with clay. In July 1984, NRC and the State of Michigan concluded that the thorium contamination exceeded the Option 1 level of the 1981 BTP. S.C. Holdings is licensed to possess 40 metric tons of thorium and 5 metric tons of uranium. The licensee completed site characterization in 1996. The buried thorium wastes were not located. There are hazardous wastes present at the site and the site is being regulated under the State of Michigan superfund law. After the radiological survey, the licensee undertook cap repair measures at the site to isolate and prevent the migration of the non-radiological hazardous wastes. A DP dated November 2003 was submitted, and on March 13, 2006 the license was amended to incorporate the DP. The licensee is requesting unrestricted release of the site. There are no immediate radiological hazards at the site. The staff has not identified any major off-site environmental issued that will not be addressed during decommissioning of the facility. The estimated cost of decommissioning the site is $1.9 million.

## 3.0 Major Technical or Regulatory Issues

The licensee under took cap repair measures at the site to isolate and prevent the migration of the non-radiological hazardous wastes. The mixture of non-radiological hazardous and radioactive waste would make the wastes unacceptable at a chemical or radioactive waste disposal site (other than an authorized mixed-waste disposal facility), and agreed to implement a monitoring program. Remediation of the site will require coordination with the Michigan Department of Environmental Quality (MDEQ), which regulates hazardous chemicals. Currently, the State of Michigan does not want the clay cap over the wastes to be removed, because of the non-radiological hazards of the site. There is minimal, if any, public interest to date. Public interest is expected to remain minimal if the clay cap is not removed. The licensee has selected unrestricted release. The probability for a hearing is low if the licensee satisfies the unrestricted release criteria with minimal disturbance to the clay cap. The potential for a hearing increases if the licensee has to remediate the site involving removal of the clay cap.

No financial assurance issues have been identified to date.

## 4.0 Estimated Date For Closure

11/01/2006

# Shieldalloy Metallurgical Corporation

## 1.0 Site Identification

Location:          Newfield, NJ
License No.:       SMB-1507
Docket No.:        04007102
License Status:    Possession-only
Project Manager:   Ken Kalman

## 2.0 Site Status Summary

The Shieldalloy Metallurgical Corporation (SMC) site is located in Newfield, New Jersey. Contamination is in the form of facility-generated slag and baghouse dust. The major contaminants are natural uranium and natural thorium. The site is also on the NPL under CERCLA, because of past operations involving chromium-contaminated onsite groundwater. In August 2001, SMC notified NRC that it had ceased production activities using source material. On August 27, 2001, the licensee provided notification and intent to decommission. The license is in timely renewal, and was amended on November 4, 2002, to authorize only decommissioning activities that were previously permitted. The licensee submitted a revised license renewal application on May 1, 2003. The licensee estimates the cost of decommissioning to be approximately 1.8 million dollars.

The SMC facility manufactures or has manufactured specialty steel and super alloy additives, primary aluminum master alloys, metal carbides, powdered metals, and optical surfacing products. One of the raw materials that was used in its manufacturing processes from 1955 to 1998 is classified as "source material" under 10 CFR Part 40. This material, called pyrochlore, is a concentrated niobium ore containing greater than 0.05 percent natural uranium and natural thorium. SMC was licensed by the NRC to ship, receive, possess, use and store source material under SMB-743. During the manufacturing process, the facility generated slag, and baghouse dust. Currently, there is approximately 18,000 m3 (635,580 ft3) of slag and approximately 15,000 m3 (529,650 ft3) of baghouse dust contaminated with natural uranium, thorium, and daughters stored on-site. Should SMC find a buyer for both the slag, which could be used as a fluidizer by steel manufacturers, and for the baghouse dust, which could be substituted for lime in the production of cement, the volume of waste would be greatly reduced, and the licensee would most likely request license termination for unrestricted use. SMC submitted a DP on August 30, 2002, which was rejected by NRC staff, because of deficiencies. SMC submitted a revised DP in October 2005, which the NRC rejected in January 2006. The staff met with SMC in March 2006 to discuss the deficiencies in the DP and develop a path forward for submittal of an acceptable DP. The NRC staff and New Jersey Department of Environmental Protection (NJDEP) staff visited the site in April 2006 to discuss erosion control design. Pursuant to comments received at these interactions, SMC submitted a supplement to its DP in June 2006. The NRC staff completed its acceptance review of the supplement in September 2006 and determined that there is sufficient information to proceed with its technical review.

## 3.0 Major Technical or Regulatory Issues

SMC has found it difficult to sell the slag material. Several attempts to export the material have failed. SMC intended to sell the baghouse dust to a local cement manufacturer, however, no buyer has been found. Regardless of whether the sales occur, SMC has proposed to dispose of these materials on-site in an engineered cell. Although the LTC approach is in the early stages of planning, the State of New Jersey has expressed concerns with the use of NRC's LTC license for the SMC site. Their concerns are: 1) the proposed approach would create an unlicensed low-level radioactive waste disposal facility; 2) that there has not been a meaningful opportunity for community discussion; and 3) the radioactive material should be disposed of and not left for future generations. SMC has less than adequate financial assurance for decommissioning.

## 4.0 Estimated Date For Closure

09/01/2010

# Stepan Chemical Company

## 1.0 Site Identification

Location:        Maywood, NJ
License No.:     STC-1333
Docket No.:     40-8610
License Status:  Possession-Only
Project Manager: Amir Kouhestani

## 2.0 Site Status Summary

The Stepan Chemical Company (Stepan) site is located in the Borrough of Maywood, NJ. Principal radioactive contaminants at the site are process wastes from the thorium extracted from the monazite sands using a chemical separation process. The residual alkaline thorium phosphate tailings are stored in three licensed underground storage areas. The license is in timely renewal. The cleanup of the licensed burials is part of a ROD prepared by USACE for soils and buildings at the FUSRAP Maywood superfund site (August 2003). Site cleanup will be guided by the USACE-NRC MOU (July 2001). In November 2004, Stepan and the United States reached a settlement on their respective obligations to remediate the site. In December 2004, Stepan informed the NRC of their views on how the site remediation will be addressed. In September 2005, the NRC and NJDEP conducted a coordinated safety inspection of the site. The MOU and the ROD commit USACE to clean up the licensed burials to meet at least the NRC standards required under 10 CFR 20.1402; Radiological criteria for unrestricted use, or a more stringent criteria.

Since Stepan's acquisition of the Maywood Chemical Works (MCW) in 1959, Stepan has been manufacturing specialty chemicals and other products at the Maywood facility. In late 1960s, Stepan conducted some site cleanup on the originial MCW plant site property on both east and west sides of the State Route 17. In accordance with the NRC regulations at the time (NRC subsequently rescinded the regulation), the waste materials were relocated to three burial areas on property currently owned by Stepan.

## 3.0 Major Technical or Regulatory Issues

The decommissioning schedule is in major part contingent upon implementation of the November 2004 settlement between the United States and the Stepan Company, the FUSRAP Appropriations, and USACE's greater FUSRAP Maywood Superfund project schedule and funding priorities. The November 2004 settlement between Stepan and the United States provides for the decommissioning financial obligations of the parties. The staff has not identified any major offsite environmental issues that will not be addressed during decommissioning of the facility. On September 29, 2005, NRC conducted a site safety inspection accompanied by the NRC project manager and an NJDEP representative. Since April 2005, USACE has initiated CERCLA site remediation in a non-licensed area of the site. On October 20, 2005, Stepan wrote to NRC and advised of their and USACE's site remediation scheduled activities. Staff is working with the licensee and USACE in order to initiate the license abeyance process consistent with the NRC-USACE MOU. Staff evaluated a December 2005, USACE request for an NRC 10 CFR 20.2002 authorization to release very low activities residuals offsite at a RCRA Subtitle C waste disposal facility in Grandview, Idaho. In April 2006,

NRC determined that USACE is not eligible for a 20.2002 authorization since USACE is not an NRC licensee or an applicant. However, the NRC noted that Stepan may request such an authorization since the license is not yet in abeyance. The NRC, licensee, and USACE further examined this option and concluded NRC does not have the statutory authority to grant such a request by the licensee. During an August 2006, management meeting between NRC and USACE, other alternatives were discussed. USACE continues to ship waste from the Maywood site and has implemented alternative approaches to its 20.2002 request to NRC.

## 4.0 Estimated Date For Closure

09/01/2009

# Superbolt (Formerly Superior Steel)

## 1.0 Site Identification

Location:              Carnegie, PA
License No.:           Non licensed facility
Docket No.:            Non licensed facility
License Status:        Unlicensed
Project Manager:       Robert Prince

## 2.0 Site Status Summary

Superbolt is the current owner of the facility.  The site was owned and operated by the Superior Steel Company, during the period from 1952 to 1957.  During this period Superior Steel performed contract work for the AEC.  Superior Steel's license expired in 1958.  The site consists of five interconnected warehouse buildings (designated as Building 23) with uranium contaminated building surfaces.  Uranium contamination is also present in a sub-floor trench located within two of the warehouse buildings and extending approximately 50 feet outside the building structure.  Historical surveys indicated the presence of ground water intrusion in portions of a sub-floor trench.  However, no indication of ground water contamination beyond the trench boundary has been detected.  Uranium contamination was also detected outside and adjacent to the building.  Currently no DP exists for the site.  The site owner does not have a cost estimate for decommissioning.

## 3.0 Major Technical or Regulatory Issues

Funding of remediation work is the primary concern.  Superbolt is a small company with limited financial resources.  Net cash flow has been limited over the last several years.  Late summer 2004, the industrial complex where the Superbolt facilities are located sustained extensive flood damage.  This situation has aggravated Superbolt's financial situation and has resulted in uninsured recovery expenses in excess of one million dollars.  Recovery efforts are still ongoing.  Superbolt officials have held discussions with DOE representatives to evaluate the original basis for excluding the Superbolt facility from FUSRAP.  In early 2006, DOE requested that USACE re-evaluate the Superbolt facility for inclusion under the FUSRAP program.  In late August 2006, USACE representatives conducted an onsite visit and are currently developing a performance assessment (PA).  The PA will address recommendations concerning whether or not the site should be included in the FUSRAP program.  Superbolt expects to retain ownership of the facilities and has no plans at this time to sell the property.

## 4.0 Estimated Date For Closure

TBD

# UNC Naval Products

## 1.0 Site Identification

Location:          New Haven, CT
License No.:       SNM-368
Docket No.:        070-00371
License Status:    Terminated
Project Manager:   Laurie Kauffman

## 2.0 Site Status Summary

This site had been used by United Nuclear Corporation (UNC) to fabricate nuclear fuel components for the U.S. Government, was decommissioned in 1976, and removed from NRC License SNM-368 on April 22, 1976. Independent measurements conducted by NRC in May 1996, and September 1996 indicated residual enriched uranium exceeding 30 pCi/g in soil or sediments in two buildings, and in a connected inactive sewer system. In June 1998, the licensee agreed to characterize and remediate the facility in accordance with Option 1 delineated in the NRC BTP for Disposal or Onsite Storage of Thorium or Uranium Wastes from Past Operations. The licensee submitted a characterization plan and DP in August 1999, and conducted sampling activities in 2003.

The UNC facility was operated by Olin Matheson Chemical Corporation from April 1956 to May 1961 and UNC from June 1961 to April 1976. The licensee's were authorized to use radioactive materials for manufacturing fuel for the naval reactor program at the site. In 1974, UNC announced the closing of the facility and transferred the inventory from this site to the Montville, CT location on their NRC license. FSSs of the New Haven facility were performed and the report submitted on February 26, 1976. NRC performed confirmatory surveys in March and October, 1976. On April 22, 1976, NRC amended SNM-368 to remove the New Haven, CT facility from the license. On June 8, 1994, license SNM-368 was terminated and the NRC released the facility for unrestricted use.

## 3.0 Major Technical or Regulatory Issues

The radioactive material on-site is not readily available and the dose consequence to the public is very low. After a review of existing contracts, DOE accepted financial responsibility for site cleanup. Site radiological activities have to be contracted through a competitive contract process that has caused some delays. UNC is not an NRC licensee and is not the current owner of the site and therefore is not required to comply with the Decommissioning Timeliness Rule, however, UNC, has agreed to undertake the remediation. The State of Connecticut and the City of New Haven have some interest in the site, as it is part of a redevelopment area and recently, the City of New Haven became the site owner through a foreclosure action. The sewer authority has cooperated with NRC and UNC in collection of sewer samples. UNC will need to obtain an access agreement with the City of New Haven before additional remediation work can be completed.

## 4.0 Estimated Date For Closure

10/07

# West Valley

## 1.0 Site Identification

| | |
|---|---|
| Location: | West Valley, NY |
| License No.: | CSF-1 |
| Docket No.: | 0500201, POOM-032 |
| License Status: | In abeyance |
| Project Manager: | Chad Glenn |

## 2.0 Site Status Summary

The West Valley site is located on the Western New York Nuclear Service Center (Center) and comprises 3,300 acres of land established for siting a former reprocessing facility. The New York State Energy Research and Development Authority (NYSERDA) holds title to this land. In its regulatory responsibilities under the Atomic Energy Act, the AEC, and subsequently NRC, licensed (CSF-1) the site from 1966 to 1981. The Center contains a former nuclear fuel reprocessing facility that operated from 1966 to 1972, and produced approximately 600,000 gallons of liquid high level waste (HLW). The Center also contains contaminated structures and two radioactive waste disposal areas: (1) a 15-acre New York State-licensed disposal area (SDA) that operated as a commercial LLW disposal facility from 1963 to 1975; and (2) a 5-acre NRC-licensed disposal area (NDA) that received radioactive wastes from the reprocessing plant and associated facilities from 1966 through 1986. In addition to the reprocessing facility and disposal areas, the Center includes a HLW tank farm, waste lagoons, above-ground radioactive waste storage areas, with soil and groundwater contamination near these facilities. In 1980, Congress enacted the West Valley Demonstration Project (WVDP) Act. Under the Act, DOE assumed exclusive possession of the 200-acre portion of the Center which includes the former reprocessing facility, the NDA, the HLW tanks, waste lagoons, and above-ground waste storage areas. The WVDP Act authorized DOE to: solidify, transport and dispose of HLW that exists at the site; dispose of LLW and transuranic waste produced by the WVDP; and decontaminate and decommission facilities used for the WVDP in accordance with requirements prescribed by NRC. In 1981, NRC put the license in abeyance to allow DOE to carry out the WVDP Act. In 2002, DOE completed the solidification of liquid HLW which was placed into 275 stainless steel canisters. The HLW canisters are expected to be stored onsite until shipped for disposal to the federal repository. In 2002, the Commission issued its final policy statement on decommissioning criteria for the WVDP. The policy statement prescribed the LTR as the decommissioning criteria for the WVDP, reflecting the fact that the applicable goal for the entire NRC-licensed site is compliance with the requirements of the LTR. DOE and NYSERDA are developing a Decommissioning and Long-term Stewardship environmental impact statement (EIS). NRC is a cooperating agency for this EIS in accordance with its responsibilities under the WVDP Act. NRC intends to use this EIS to fulfill its National Environmental Policy Act (NEPA) responsibilities for applying the LTR to the WVDP and to assist in its determination of whether the preferred alternative meets the LTR. In September 2005, NRC and other cooperating agencies initiated a review of a preliminary (pre-decisional) draft of this EIS. In April 2006, NRC and other cooperating agencies forwarded comments to DOE on this preliminary draft EIS. DOE's current schedule forecasts public release of the draft EIS in 2007, and release of the final EIS in 2008. DOE also plans to submit a DP for NRC review in 2007.

## 3.0 Major Technical or Regulatory Issues

- Long-term site stewardship

- Waste Incidental to Reprocessing

- Effects of erosion on disposal areas

- Groundwater contamination

- Payment of HLW disposal fees

## 4.0 Estimated Date For Closure

TBD

# Westinghouse Electric Company (Churchill Facility)

## 1.0 Site Identification

Location:          Pittsburgh, PA
License No.:       SNM-1460
Docket No.:        07001503
License Status:    Active
Project Manager:   Dave Everhart

## 2.0 Site Status Summary

Licensee uses byproduct and SNM radioactive materials for research and development related to commercial nuclear power reactors. The Facility is situated on 148 acres, and consists of ten major buildings (70,000 square feet) with office space and laboratories.

License No. SNM-1460 was issued on 1955, pursuant to 10 CFR Part 70, and has been amended periodically since that time. This license authorized the licensee to use any byproduct material with atomic numbers from 1 through 96, plus Californium-252, sealed and unsealed for purposes of conducting research and development activities on laboratory bench tops and in hoods. Westinghouse submitted a DP in May 2005. The staff is currently reviewing the DP.

## 3.0 Major Technical or Regulatory Issues

None.

## 4.0 Estimated Date For Closure

TBD

# Westinghouse Electric Company (Hematite Facility)

## 1.0 Site Identification

Location: Festus Township, Jefferson County, MO
License No.: SNM-33
Docket No.: 07000036
License Status: Active
Project Manager: Amy Snyder

## 2.0 Site Status Summary

The property consists of approximately 228 acres. The operating facility consists of two main plant buildings, an administration and several support buildings, and a parking area. Plant operations included low-enriched uranium fuel fabrication, processing and treating uranium compounds, including all forms of uranium from depleted to enriched uranium, and thorium fuel. Contamination at the site consists of uranium and thorium in the soil and groundwater. The Westinghouse Electric Company, LLC (WEC) has provided phased-notification of cessation of operational activities. On September 11, 2001, WEC provided notification of cessation of all principal activities and submitted an application for license amendment to change the scope of authorized license activities to those associated with decommissioning activities. WEC has performed, within its permitted license activities, certain equipment decontamination and dismantlement and has shipped equipment and material to its South Carolina facility. NRC has determined that one EA is required to avoid segmentation under NEPA. Decommissioning is estimated to cost approximately $40.5 million.

Throughout its history, Hematite facility's primary function has been to manufacture uranium metal and uranium compounds from natural and enriched uranium for use as nuclear fuel. From its inception in 1956 through 1974, the facility was used primarily in support of Government contracts that required production of highly enriched uranium products. From 1974 through the plant closure in 2001, the focus changed from government contracts to commercial fuel production plant. Over the lifetime of the facility there have been six owners. Mallinckrodt, United Nuclear and Gulf United Nuclear owned the plant for the government focused phase of operations. Combustion Engineering, ABB and Westinghouse owned the plant during the commercial phase of operations.

## 3.0 Major Technical or Regulatory Issues

WEC submitted its comprehensive DP on October 5, 2005. WEC requested that NRC approve an alternate schedule. NRC notified WEC that it will not accept the DP for a detailed technical review at this time due to technical reasons. The licensee resubmitted the DP in June 2006. Based on the review of the DP, the following key technical issues were identified: Site characterization of the burial pits and groundwater is insufficient; technical basis documents that support the FSS design are not detailed enough or need to be developed; justification for dose modeling scenario and parameters needs to be provided; the cost estimate is not detailed enough for a detailed evaluation; and there is a general lack of information necessary for staff to complete its EA. WEC met with NRC staff on January18, 2006, in a pre-licensing meeting, to discuss proposed criticality amendment and request for security exemptions related to burial pit remediation. Although, only one EA and one DP will be produced, Westinghouse still has plans to address NRC regulatory requirements concurrent with those required under EPA's CERCLA process. This coordination of remedial investigation and remedial action under CERCLA versus NRC's LTR decommissioning and site cleanup criteria could potentially be challenging. There are active local, State, and Congressional interests in how the site will be decommissioned. WEC has sued previous owners and the US government for cost recovery for decommissioning. Westinghouse has entered into an agreement with the State of Missouri to give the State specific authority in the decommissioning of the site. NRC has submitted comments to the State of Missouri opposing the consent agreement because it does not recognize NRC authority over radioactive cleanup.

## 4.0 Estimated Date For Closure

03/01/2012

# Westinghouse Electric Company (Waltz Mill)

## 1.0 Site Identification

Location:            Madison, PA
License No.:         SNM-770
Docket No.:          070-00698
License Status:      Active
Project Manager:     Mark Roberts

## 2.0 Site Status Summary

The WEC Waltz Mill site is currently licensed primarily to provide testing, calibration, and maintenance services for contaminated reactor servicing equipment and other reactor components. Radiological contamination in soil and groundwater existed on a portion of the site as a result of the clean-up activities following a 1961 incident at the TR-2 test reactor, waste segregation activities, and nuclear laundry services. Significant contamination was also present in retired facilities (hot cells, hot cell support rooms, and a section of the fuel transfer canal) within one of the site buildings. Contaminants are primarily Sr-90 and Cs-137, with lesser quantities of mixed fission, activation products, and trace levels of transuranic radionuclides. Due to a series of corporate mergers, the licensee for the TR-2 test reactor is CBS, Inc. WEC submitted a RP in April 1997. NRC approved the RP in January 2000. The licensee has remediated much of the interior and exterior contaminated areas. Contaminated soil removal has been completed in the primary exterior contaminated area, although small pockets of contaminated soil and a major contaminated process drain line remain on the site. WEC and CBS have over $40 million in financial assurance agreements in place for completion of site decommissioning.

## 3.0 Major Technical or Regulatory Issues

The licensee does not intend to request termination of the license. The licensee and CBS went forward with the remediation project, in part, to address the reasons why the facility was originally placed on the SDMP list. The TR-2 license was intended to be terminated after decommissioning of the test reactor and transfer of the building to the WEC SNM-770 license. WEC and CBS have not reached an agreement on the transfer because of a disagreement on the completion status of remediation activities. This and related issues are currently being resolved under arbitration. PADEP has interest in the condition of the site, particularly groundwater issues. No financial assurance issues have been identified at this time.

## 4.0 Estimated Date For Closure

10/07

# Whittaker Corporation

## 1.0 Site Identification

Location:          Greenville, PA
License No.:       SMA-1018
Docket No.:        040-7455
License Status:    Possession-only
Project Manager:   James Kottan

## 2.0 Site Status Summary

The Whittaker Corporation (Whittaker) site is located within an industrial park, approximately 6 km south of Greenville, PA. The site comprises a 5.9 acre strip of land located between the Greenville Metals Plant and the Shenango River. The site is divided into four sections. Section 1 includes the southern end of the site and consists of a mixture of slag and gravel which sits above a tributary leading to the Shenango River. Metal scraps are observed within the slag and gravel mixture and the northern end of the section. No large pieces of slag or elevated readings have been observed in Section 1. Sections 2 and 4 are located in the center of the site. This area is comprised predominately of slag material. Two visually distinct types of slag are present. One slag is blue-green and the other is black. The blue-green slag has a glassy texture and the black slag is porous and rocklike. The black slag contains the radioactive material. Section 3 comprises the northern end of the site. A large part of Section 3 is covered by a concrete slab. Three sided bins containing slag material and piles of slag mixed with other debris are on top of the concrete pad. The bins contain low-level waste source materials and non-toxic industrial waste some of which is also contained in rusting drums. The eastern portion of the Section 3 is densely vegetated. Facility topography (prior to the initiation of decommissioning) had been built up through the repeated disposal of slag, scrap metal, debris, and foundry sand. The slag piles had reached elevations of twenty feet or more above the adjoining river flood plain. The slag piles in Section 2 have been excavated and screened to remove the radioactive material, which was shipped for disposal.

Mercer Alloys Corporation was founded in 1955 for the purpose of reclaiming valuable scrap metals from old jet engines and aircraft. The operations were later expanded to include processing ferro-columbium, ferro-nickel, and ferro-molybdenum alloys from ores, as well as accepting other forms of scrap metal. Some of the raw materials and feedstock used in these processes contained licensable quantities of natural thorium or uranium. The AEC issued License SUB-864, to Mercer Alloys in February 1966, for the possession of 250 pounds of uranium. The company was purchased by Whittaker Corporation in 1967, and the license was allowed to expire in February 1969, because no radioactive materials had been procured or used. However, the licensee began receiving columbium ore containing source material (thorium-232) in October 1969, prompting them to apply for a new license. AEC issued License SMA-1018, to Mercer Alloys (Whittaker) on December 15, 1969, for the possession of 16,000 pounds of source material. Processing of the columbium ores resulted in concentrating thorium, and some of the processed scrap metals contained natural and depleted uranium. Both of these contaminants were concentrated and retained in the resulting waste slag. Processing operations utilizing licensable materials ceased in 1974, and Whittaker sold the metal alloys division to Exomet, Inc. However, Whittaker maintained ownership and responsibility for the source material. In 1975, Whittaker initiated decontamination of the equipment and plant areas. Contaminated equipment, rubble, and slag resulting from these cleanup efforts were added to existing slag and waste piles located in the site's eastern section. The portion of the property housing the plant was released for unrestricted use in 1975. An additional plant building was decommissioned in 1983 and released for unrestricted use in 1985. This plant side of the property remains an active business, now operated by Greenville Metals, and is not associated with Whittaker or the remaining licensed area. The plant is separated from the slag and waste site by metal fencing. Thorium-and-uranium bearing wastes, raw materials, feed-metal scrap, and contaminated building materials that were generated from the facility decontamination activities are contained in the licensed and controlled waste and slag storage areas. In 2004, the site initiated decommissioning activities, starting in Section 2, where the highest activity slag was believed to be located.

## 3.0 Major Technical or Regulatory Issues

During excavation of Section 2, additional subsurface contamination was identified that extends beyond the fence separating the property from the Greenville Metals site. The material is below ground level, and may not be accessed from the uncontrolled side of the fence. There are no adverse safety consequences to the public or to workers at the Greenville Metals site due to this material. Its discovery, however, will require Whittaker to characterize how far onto the property the material extends. In addition to the subsurface material, in May 2006, surface pieces of contaminated slag were discovered on the Greenville Metals property, as well as subsurface slag in previously-unidentified locations. Pieces that could be carried by hand were removed and relocated to the Whittaker site. The plant owners were notified of the presence of the material and instructed to not remove it. The material does not represent a health or safety risk to the public. It does not meet the activity levels requiring posting or control. The Greenville Metals site is surrounded by a fence, which provides de facto control over the material. Whittaker and Greenville are developing an agreement to allow Whittaker to remediate any slag from this site. On the Whittaker property, contamination was identified at depths that are deeper than had been expected based on characterization data. The slag pile was expected to reach between 15 - 20 feet below grade. In one location, contamination reaches approximately 25 feet. The material is being excavated and removed in the same manner as the previously-removed material.

## 4.0 Estimated Date For Closure

01/31/2008

# Appendix D

# Site Summaries for Decommissioning Title II Sites

# American Nuclear Corporation

## 1.0 Site Identification

Location:          Casper, WY
License No.:       SUA-667
Docket No.:        40-4492
License Status:    Possession-only
Project Manager:   Myron Fliegel

## 2.0 Site Status Summary

Reclamation oversight of the facility has been transferred to the State of Wyoming, Department of Environmental Quality (WDEQ). This transfer occurred because American Nuclear Corporation (ANC) had become insolvent in May 1994 and site reclamation was incomplete. A Confirmatory Order between the U.S. Nuclear Regulatory Commission (NRC) and the WDEQ describing the requirements for reclamation activities was agreed upon by both parties and was issued in October 1996.

The licensed site encompasses approximately 550 acres of which approximately 80 acres consist of Tailings Pile 2, and 40 acres of Tailings Pile 1. Tailings Pile 2 reclamation activities were completed and approved by NRC in February 1998. Tailings Pile 1 activities are on hold. Additionally, the site has an active groundwater recovery and corrective action program.

Reclamation activities were targeted to restart in 2005, but have not. Since the last inspection in September 2005, no site reclamation activities had been performed. After approval of the reclamation plan for Tailings Pile 1, activities will include the following: (1) windblown area clean-up activities, (2) capping with clay, (3) radon testing, and (4) placement of rip-rap rock. WDEQ's goal was to submit the final reclamation plan for Tailings Pile 1 for NRC review and approval in 2004 and complete the associated reclamation in 2005, but that has not been accomplished. The cost for decommissioning is estimated to be approximately $3.2 million. WDEQ has approximately $3.2 million in DOE Title10 funds to complete reclamation of Pile 1.

## 3.0 Major Technical or Regulatory Issues

None.

## 4.0 Estimated Date For Closure

TBD

# Bear Creek

## 1.0 Site Identification

Location: Converse County, WY
License No.: SUA-1310
Docket No.: 40-8452
License Status: Possession-only
Project Manager: Myron Fliegel

## 2.0 Site Status Summary

The decommissioning and reclamation of the Bear Creek Uranium Mill, including the mill tailings impoundment, was completed in November 1999. The tailings impoundment contains 4.7 million tons of uranium ore tailings and covers an area of approximately 101 acres. The staff performed a final "closeout" inspection of the completed reclamation construction activities in July 2000. The staff completed its review of the Bear Creek Tailings Reclamation Construction Report in July 2001, with the conclusion that reclamation of the Bear Creek tailings impoundment was completed in accordance the requirements of 10 CFR Part 40, Appendix A, and the licensee's approved Tailings Reclamation Plan. At the Bear Creek site, the State of Wyoming owns both the surface estate (where the tailings impoundment is located) and the subsurface estate with the contained mineral rights. The licensee purchased the surface estate from the State in January 2003 and is currently negotiating with the State to acquire the subsurface estate. Following the successful acquisition of the subsurface estate, the licensee can prepare the necessary papers to turn over ownership of the site to the U.S Department of Energy (DOE) for long-term custody and subsequent termination of its NRC license. The cost for decommissioning is estimated to be approximately $900,000.

## 3.0 Major Technical or Regulatory Issues

There is one major regulatory issue at the Bear Creek site that must be addressed as part of the license termination process. That issue relates to the current State ownership of the subsurface estate at the site. If the State does not divest itself of ownership of the subsurface estate, the State will become the long-term custodian for those interests. However, the State has previously expressed its disinterest in becoming the long-term custodian for any Title II (Uranium Mill Tailings Radiation Control Act [UMTRCA] of 1978, as amended) site in Wyoming. Accordingly, licensee and State representatives are currently negotiating an exchange of mineral interests (the subsurface estate) so that the licensee can acquire ownership of the entire Bear Creek site, including the subsurface interests. License termination depends on: (a) a successful exchange of mineral estates between the licensee and the State; (b) transfer of the Bear Creek site property by the licensee to the DOE; (c) payment of the long-term surveillance fee by the licensee; and (d) NRC approval of the Long-term surveillance plan (LTSP) currently being prepared by the DOE.

## 4.0 Estimated Date For Closure

2007

# Cogema Mining, Inc.

## 1.0 Site Identification

Location:            Mills, WY
License No.:         SUA-1341
Docket No.:          40-8502
License Status:      Possesson-only
Project Manager:     Ron Linton

## 2.0 Site Status Summary

This site includes the Irigaray and Christensen Ranch in situ leach mines in one license, NRC License SUA-1341. The area of these two in-situ leach (ISL) sites (7 miles apart) disturbed by well fields or facilities is approximately 687 acres. COGEMA's license was changed to a possession-only license in 2001 after active in-situ leach mining ceased and groundwater restoration commenced. Groundwater restoration is complete at Irigaray and nearly complete at Christensen Ranch. COGEMA submitted a restoration report for Irigaray Mine units 1 through 9 to NRC staff in November 2005, as well as to WDEQ for approval of ground water restoration at the Mine. NRC staff concurred with WDEQ's approval that the ground water had been restored to pre-mining class of use. Surface decommissioning began in 2002 (plan approved December 31, 2001), and there are minor amounts of soil contamination. Equipment and building components that cannot be decontaminated will be shipped with the contaminated soil to the Pathfinder - Shirley Basin tailings impoundment for disposal. The cost for decommissioning is estimated to be approximately $12.1 million.

## 3.0 Major Technical or Regulatory Issues

None.

## 4.0 Estimated Date For Closure

TBD

# Exxonmobil Highlands

## 1.0 Site Identification

Location:        Converse County, WY
License No.:     SUA-1139
Docket No.:      40-8102
License Status:  Possession-only
Project Manager: Myron Fliegel

## 2.0 Site Status Summary

The Highland uranium recovery facility included a conventional surface uranium mine with an associated mill. The site also included ore storage pads, four mine pits (two of which have been backfilled), several waste rock piles, one tailings impoundment and on environmental laboratory. Surface mining, solution mining and underground mining were used to recover the uranium ore. The first ore was processed at the Highland mill using an acid leach circuit between 1972 and the end of operations in mid-1984.

The uranium mill area, including the ore storage pads and the laboratory, has been cleaned up and the tailings buried are under a radon barrier, eliminating nearly all potential for radiation exposures to workers or members of the general public from these sources. However, Exxon Mobil reclamation operations may result in minimal exposure of workers to radioactive materials. All windblown material has been reclaimed to unrestricted release standards. The byproduct material exposure is limited to the groundwater pathway. However, there is no current use of groundwater.

## 3.0 Major Technical or Regulatory Issues

None.

## 4.0 Estimated Date For Closure

2008

# Homestake

## 1.0 Site Identification

Location:          Grants, NM
License No.:       SUA-1471
Docket No.:        40-8903
License Status:    Possession-only
Project Manager:   Ron Linton

## 2.0 Site Status Summary

The facility is a conventional uranium mill site under reclamation. Uranium processing started in the late 1950's and continued until 1990. Tailings generated from the milling operation were placed on two piles, a large pile and a small pile. The facility has a tailings area of 170 acres with a weight of 22 million tons. Currently there are several evaporation ponds and an ion exchange treatment building for groundwater remediation, and several administrative and maintenance buildings. Seepage from the tailings piles was noted in 1975.

The current effort is a major groundwater corrective action plan which is also under the oversight of the U.S. Environmental Protection Agency (EPA) through Superfund. A memorandum of understanding (MOU) has been executed between NRC and EPA for this site regarding groundwater remediation. Staff has recently completed a license amendment to revise groundwater compliance standards. In accordance with the MOU, staff is consulting with EPA and the State of New Mexico to review this action. The cost for decommissioning is estimated to be approximately $55.5 million and is expected to be completed by 2017.

## 3.0 Major Technical or Regulatory Issues

None.

## 4.0 Estimated Date For Closure

2017

# Pathfinder - Lucky MC

## 1.0 Site Identification

Location:           Gas Hills Mining District, WY
License No.:        SUA-672
Docket No.:         40-2259
License Status:     Possession-only
Project Manager:    Steve Cohen

## 2.0 Site Status Summary

The Lucky Mc site is located in west central Wyoming in the Gas Hills region. Uranium milling began at this site in 1958 and continued through 1988 with a total of 12 million tons of ore processed. The mill utilized a conventional acid leach process. The mill was demolished and placed in the out-slope of the No. 2 Tailings Dam, with a clay-radon barrier placed over the material. The mill area includes approximately 56 acres. The site has three solid tailings impoundments and three tailings solution ponds. The post-reclamation tailings piles cover approximately 241 acres.

Ground water pumping operations at the facility have been ongoing since 1980. The corrective action consists of ground water pumping to evaporation ponds and the injection of fresh water to remove contamination and impede the flow of contaminated ground water in the aquifer. A total of 197 million gallons of contaminated water has been collected and 193 million gallons of fresh water injected as part of the remedial effort and approximately 217 million gallons of water have been pumped from the tailings by the end of 2001. On December 20, 2002, alternate concentration limits (ACLs) were approved for the Lucky Mc site, and all active correction actions ceased, such as pumping and injection. The mill tailings site reclamation was completed on December 14, 2004. The Construction Completion Report was submitted to the NRC on April 21, 2005. Staff review of the document is estimated to be completed by 2006. The cost for decommissioning is estimated to be approximately $1.0 million.

## 3.0 Major Technical or Regulatory Issues

None.

## 4.0 Estimated Date For Closure

2007

# Pathfinder - Shirley Basin

## 1.0 Site Identification

| | |
|---|---|
| Location: | Shirley Basin, WY |
| License No.: | SUA-442 |
| Docket No.: | 40-6622 |
| License Status: | Active |
| Project Manager: | Steve Cohen |

## 2.0 Site Status Summary

The Shirley Basin site is located in eastern Wyoming. The former uranium mill and mine site is located approximately 5 miles northeast of the former Shirley Basin town site. Uranium milling began on the site in 1971, and continued through 1992, when the last ore was processed. The site has two solid tailings impoundments, the largest covering approximately 158 acres, and the smaller 135 acres. A solution pond, which is also the disposal location for 11e.2 byproduct material from ISL mines, covers about 30 acres. The mill has been decommissioned in accordance with the decommissioning plan (DP) submitted to NRC in 1992. Pathfinder has reclaimed approximately 80 percent of the tailings and has discontinued groundwater reclamation, due to staff's approval of the pro[posed ACLs  pathfinder intends to operate its ISL disposal area for the forseeable future.

## 3.0 Major Technical or Regulatory Issues

Staff issued the license amendment for ACLs in October 2005. To address concerns regarding sudden pollutant loads to Spring Creek, a license amendment was added requiring Pathfinder to maintain its ground water remediation system until the staff determines that it is no longer needed.

## 4.0 Estimated Date For Closure

TBD

# Rio Algom - Ambrosia Lake

## 1.0 Site Identification

Location:      Grants, NM
License No.:    SUA-1473
Docket No.:    40-8905
License Status:  Possession-only
Project Manager:  Robert Lukes

## 2.0 Site Status Summary

This is a uranium mill tailings site in the Ambrosia Lake uranium district of New Mexico. It is located approximately 25 miles north of Grants, New Mexico. The tailings impo8undment contains 33 million tons of uranium ore and covers an area of approximately 370 acres.

The site status changed from standby to reclamation in August 2003 to reflect the licensee's intent to begin full demolition and reclamation of the site leading to termination of the specific license. The mill was demolished and disposed of in the tailings impoundment in late 2003. The demolition was completed in accordance with a mill demolition plan approved by NRC in October 2003.

The staff issued a license amendment for ACLs in February 2006. Consequently, all groundwater corrective actions have been discontinued, and rio Algom is finalizing the site tailings reclamation. A portion of the tailings impoundment is still open for disposal of Atomic Energy Act, Section 11e.(2) byproduct material. A final soil DP entitled, Closure Plan - Lined Evaporation Ponds (Relocation Plan) was submitted to the NRC in November of 2004, and partially approved. A portion of the report, pertinent to the "Section 4" and Pond 9 evaporation pond sediment material is still under review. It is estimated that that portion of the review will be completed by 2006.

The cost for decommissioning is estimated to be approximately $18 million.

## 3.0 Major Technical or Regulatory Issues

None.

## 4.0 Estimated Date For Closure

2009

# Sequoyah Fuels Corporation

## 1.0 Site Identification

Location: Gore, OK
License No.: SUB-1010
Docket No.: 04008027
License Status: Possession-Only
Project Manager: Myron Fliegel

## 2.0 Site Status Summary

Uranium and thorium contamination of the soils and subsoils has been identified at the site. In addition, the groundwater is contaminated with uranium, thorium and metals. In March 1999, Sequoyah Fuels Corporation (SFC) submitted a DP to remediate the site and terminate the license in accordance with the License Termination Rule (LTR)(10 CFR 20.1403 for restricted conditions). In order for NRC to approve the DP, a long-term custodian to assume responsibility for institutional controls would have to be identified. However, several potentially acceptable custodians declined, and SFC was unable to get a commitment from DOE. In January 2001, SFC requested that some of the waste at the site be classified as 11e.(2) byproduct material and thus subject to Appendix A of Part 40. In July 2002, the NRC approved SFC's request that some of the wastes be classified as 11e.(2). In December 2002, the license was amended to permit possession of 11e.(2) and require site remediation in accordance with Appendix A of Part 40. After remediation and license termination, DOE will be required to assume responsibility, as the long-term custodian, if the State of Oklahoma chooses not to. SFC submitted a reclamation plan in January 2003. The staff is currently reviewing the plan and developing an associated EIS. In June 2003, SFC submitted a ground water monitoring plan and a ground water corrective action plan. In August 2005 the staff approved the ground water monitoring plan. The staff is currently reviewing the ground water corrective action plan.

## 3.0 Major Technical or Regulatory Issues

There is significant groundwater contamination at this site which the groundwater monitoring and corrective action plan are intended to address. Staff also has concerns with the placement and design of the waste impoundment. Financial assurance issues are summarized in SECY-03-0198 dated November 12, 2003. A hearing was granted to the State of Oklahoma and the Cherokee Nation on issues related to the reclamation plan proposed by SFC.

## 4.0 Estimated Date For Closure

TBD

# Umetco Minerals Corporation

## 1.0 Site Identification

Location:            Gas Hills, WY
License No.:         SUA-648
Docket No.:          40-0299
License Status:      Possession-only
Project Manager:     Robert Lukes

## 2.0 Site Status Summary

The site is located in the Gas Hills uranium district of the Wind River Basin. The restricted area, including the tailings disposal and heap leach areas, consists of approximately 542 acres, of which Umetco Minerals Corporation (Umetco) owns 280 acres.

Mill operation ended in 1984 and the mill was decommissioned in 1990. The final status survey (FSS) report is under NRC review, but one building in the restricted area and a small portion of the haul road have yet to be remediated. The windblown tailings remediated area is 111 acres, and 4,950 cubic yards of soil were removed (soil DP approved April 2001). An additional 6,700 cubic yards of material were removed because of contamination released when the tailings dam was breached, and 30,000 cubic yards were removed from a former evaporation pond.

The covers on two disposal areas are complete, and the cover is nearly complete on the A-9 Repository (former uranium mining pit). The area of Pond 2 will be covered next year, if enough water evaporates (cover design approved November 10, 2003), and the C-18 uranium mining pit will be backfilled. The total disposal area is approximately 300 acres. Before license termination, DOE must arrange for transfer of land that is within the long-term care boundary from the Bureau of Land Management. The cost for decommissioning is estimated to be approximately $14.8 million.

## 3.0 Major Technical or Regulatory Issues

None.

## 4.0 Estimated Date For Closure

2010

# United Nuclear Corporation

## 1.0 Site Identification

| | |
|---|---|
| Location: | Churchrock, NM |
| License No.: | SUA-1475 |
| Docket No.: | 40-8907 |
| License Status: | Possession-only |
| Project Manager: | Paul Michalak |

## 2.0 Site Status Summary

The facility is a conventional uranium mill site under reclamation. United Nuclear Corporation (UNC) operated the site as a uranium mill facility from 1977 to 1982. The site includes a former ore processing mill and tailings disposal area, which cover about 25 and 100 acres, respectively. The mill, designed to process 4,000 tons of ore per day, extracted uranium using conventional crushing, grinding, and acid-leach solvent extraction methods. Uranium ore processed at the site came from the Northeast Church Rock and the Old Church Rock mines. The average ore grade processed was approximately 0.12 percent uranium oxide. The milling of uranium ore produced an acidic slurry of ground waste rock and fluid (tailings) that was pumped to the tailings disposal area. Uranium milling and tailings disposal were conducted and an estimated 3.5 million tons of tailings were disposed in the tailings impoundments. The tailings disposal area is subdivided by dikes into three cells identified as the South Cell, Central Cell, and North Cell. Surface reclamation is complete, except for the area of the south tailings cell covered by two evaporation ponds, which are part of the ground water corrective action plan. The current effort is a groundwater corrective action plan which is also under oversight of the EPA through Superfund. A MOU has been executed between NRC and EPA for this site.

UNC is developing a Site-wide Supplemental Feasibility Study (SWSFS) as directed by EPA on June 24, 2005. The SWSFS is scheduled to be submitted for review by November 30, 2006. The cost for decommissioning is estimated to be approximately $3.7 million.

## 3.0 Major Technical or Regulatory Issues

Ground water corrective action at the UNC Church Rock site involves three saturated units; the Southwest Alluvium, Zone 1 and Zone 2. For the Southwest Alluvium, the corrective action system has been temporarily shut down while UNC assesses the effectiveness of natural attenuation as a ground water remedial solution. Quarterly ground water quality monitoring is ongoing in the Southwest Alluvium. For Zone 1, the corrective action system, which was initiated in 1984, was decommissioned in July 1999 with the approval of the NRC, EPA, and NMED. A monitored natural attenuation approach has been proposed for Zone 1. Currently, monitoring of the natural system's ability to stabilize seepage impacts into Zone 3 is continuing.

Recently, UNC conducted an extended pilot investigation to evaluate the suitability of hydrofracturing to enhance the remedy for cutoff and containment of the migrating seepage-impacted Zone 3 water. Additionally, EPA has approved a supplemental pilot study for testing in-situ alkalinity stabilization to stop further migration of the seepage-impacted Zone 3 water.

Given stakeholder interest in the Church Rock site, significant coordination with various interest groups will be necessary. The Navajo Nation has had a major interest in the site.

Sedimentation issues in the diversion channel have been communicated to the licensee as they relate to license termination.

## 4.0 Estimated Date For Closure

TBD

# Western Nuclear, Inc. - Split Rock

## 1.0 Site Identification

Location:          Jeffrey City, WY
License No.:       SUA-56
Docket No.:        40-1162
License Status:    Possession-only
Project Manager:   Steve Cohen

## 2.0 Site Status Summary

This site is a uranium mill tailings site located approximately 2 miles west of Jeffrey City, WY. The site consists of three reclaimed tailings impoundments occupying approximately 180 acres and other reclaimed disposal areas.

Mill operations commenced in 1958 and continued until 1981. Uranium ore processed at the mill was extracted in mines south of the facility. The mill operations consisted of physical and chemical including sulfuric acid leaching. Decommissioning of the mill was completed on September 15, 1988. Surface reclamation has been completed for the tailings impoundments. Two evaporation ponds were in use to support groundwater corrective action. However, ACLs were approved on September 28, 2006; therefore, Western Nuclear will discontinue groundwater corrective actions and will reclaim the two evaporation ponds.

## 3.0 Major Technical or Regulatory Issues

Staff issued the license amendment for ACLs with institutional controls (ICs). ICs will serve to prevent human exposures to contaminated groundwater within the long-term surveillance boundary. The NRC approved the use of ICs in December 2002, and the last IC was obtained in January 2006.

## 4.0 Estimated Date For Closure

2008

# Appendix E

# Site Summaries for Decommissioning Fuel Cycle Facilities

# AREVA NP

## 1.0 Site Identification

| | |
|---|---|
| Location: | Richland, WA |
| License No.: | SNM-1227 |
| Docket No.: | 70-1257 |
| License Status: | Active |
| Project Manager: | Merritt N. Baker |

## 2.0 Site Status Summary

This facility had five lagoons which were used as part of the waste-water treatment process. The Washington State Department of Ecology ("Ecology") ordered the licensee to close the lagoons. The licensee and Ecology entered into a consent decree agreement identifying enforceable milestones for completion of closure of the lagoons. These milestones include emptying the lagoons by September 8, 2004, and submitting a closure certificate for the lagoons by August 8, 2006. By 2004, the lagoons were removed from service and sealed off from process streams. All removable liquid has been removed and processed. Sludge has been removed from all the lagoons. Sub-liner soil removal is completed. The closure data package has been submitted to the State.

## 3.0 Major Technical or Regulatory Issues

Framatome changed its name to AREVA NP (ANP), effective 3/15/06. ANP's updated decommissioning funding plan was approved in June 2006, and reflects revised estimates as lagoon closure is complete.

## 4.0 Estimated Date For Closure

TBD

# General Atomics

## 1.0 Site Identification

| | |
|---|---|
| Location: | San Diego, CA |
| License No.: | SNM-696 |
| Docket No.: | 70-734 |
| License Status: | Possession-only |
| Project Manager: | Merritt Baker |

## 2.0 Site Status Summary

In September 1996 General Atomic's (GA's) special nuclear material (SNM) license was amended to authorize only decommissioning activities. By an application dated October 11, 1996, and supplements dated December 5, 1996; April 18, 1997; and January 15, 1998; GA requested an amendment to its license to incorporate a site decommissioning plan (DP). In accordance with the DP, GA is undergoing site wide decommissioning in accordance with the DP. All areas have been decommissioned, and the FSS for the last area is expected in late 2006.

GA's request to lower the possession limit to less than a critical mass was approved in 2003.

The primary radioactive contaminant is uranium-235. Soil will be remediated to levels specified in option 1 of the branch technical position (BTP), "Disposal or Onsite Storage of Thorium or Uranium Wastes from Past Operations," [46 FR 52061; October 23, 1981]. Facilities and equipment will be decontaminated to levels specified in "Guidelines for Decontamination of Facilities and Equipment Prior to Release for Unrestricted Use or Termination of Licenses for Byproduct, Source, or Special Nuclear Material," [USNRC, Policy and Guidance Directive FC 83-23, Division of Industrial and Medical Nuclear Safety, November 4, 1983]. GA intends to decommission areas required for unrestricted use and to terminate its SNM license.

## 3.0 Major Technical or Regulatory Issues

None.

## 4.0 Estimated Date For Closure

2007

# Honeywell

## 1.0 Site Identification

| | |
|---|---|
| Location: | Metropolis, IL |
| License No.: | SUB-526 |
| Docket No.: | 40-3392 |
| License Status: | Active |
| Project Manager: | Mike Raddatz |

## 2.0 Site Status Summary

This facility is the only operational conversion facility in the United States. There are two CaF2 settling ponds on this site. In 2001, NRC determined that the material in the ponds could be treated as exempt material, as defined in 10 CFR 40.13(a), and should be disposed of accordingly. In 2003, this licensee remediated "A" pond and disposed of the solid material. Treatable liquid has been removed from "C" pond, but the licensee is planning to treat the solids differently than in the past.

The Honeywell uranium conversion facility is on 1100 acre site (60 acres within the fence-line). Honeywell is authorized to convert natural uranium ore to natural uranium hexafluoride. Uranium conversion process occurs in the Feeds Material Building. Honeywell is authorized to possess 150 million pounds of natural uranium for the chemical conversion of uranium ore concentrates into uranium hexafluoride.

## 3.0 Major Technical or Regulatory Issues

None.

## 4.0 Estimated Date For Closure

TBD

NRC FORM 335
(9-2004)
NRCMD 3.7

U.S. NUCLEAR REGULATORY COMMISSION

1. REPORT NUMBER
(Assigned by NRC, Add Vol., Supp., Rev., and Addendum Numbers, if any.)

NUREG-1814, Volume 1

## BIBLIOGRAPHIC DATA SHEET

*(See instructions on the reverse)*

**2. TITLE AND SUBTITLE**

Status of the Decommissioning program
2006 Annual Report
Final Report

**3. DATE REPORT PUBLISHED**

| MONTH | YEAR |
|---|---|
| February | 2007 |

**4. FIN OR GRANT NUMBER**

**5. AUTHOR(S)**

John T. Buckley

**6. TYPE OF REPORT**

Technical

**7. PERIOD COVERED** *(Inclusive Dates)*

10/01/2005 thru 9/30/2006

**8. PERFORMING ORGANIZATION - NAME AND ADDRESS** *(If NRC, provide Division, Office or Region, U.S. Nuclear Regulatory Commission, and mailing address; if contractor, provide name and mailing address.)*

Division of Waste Management and Environmental Protection
Office of Federal and State Materials and Environmental Management Programs
U.S. Nuclear Regulatory Commission
Washington, DC 20555-0001

**9. SPONSORING ORGANIZATION - NAME AND ADDRESS** *(If NRC, type "Same as above"; if contractor, provide NRC Division, Office or Region, U.S. Nuclear Regulatory Commission, and mailing address.)*

same as above

**10. SUPPLEMENTARY NOTES**

**11. ABSTRACT** *(200 words or less)*

This report provides a comprehensive overview of the U.S. Nuclear Regulatory Commission's (NRC's) decommissioning program. Its purpose is to provide a stand-alone reference document that describes the decommissioning process and summarizes the status of decommissioning activities, under NRC jurisdiction, through September 30, 2006. This includes the decommissioning of complex decommissioning sites, commercial reactors, research and test reactors, uranium recovery facilities, and fuel cycle facilities. In addition, this report discusses accomplishments of the decommissioning program in fiscal year (FY) 2006; identifies the key decommissioning program issues that the staff will address in FY 2007; and provides information Agreement States have supplied on decommissioning in their States.

**12. KEY WORDS/DESCRIPTORS** *(List words or phrases that will assist researchers in locating the report.)*

Decommissioning
Complex Material Sites
2006 Annual Report

**13. AVAILABILITY STATEMENT**

unlimited

**14. SECURITY CLASSIFICATION**

*(This Page)*

unclassified

*(This Report)*

unclassified

**15. NUMBER OF PAGES**

**16. PRICE**

NRC FORM 335 (9-2004)